Albert Einstein

THE HUMAN SIDE

Presented to

on the occasion of

THE EINSTEIN CENTENNIAL SYMPOSIUM

March 4-9, 1979

THE INSTITUTE FOR ADVANCED STUDY

Harry Woolf, Director

Albert Einstein

THE HUMAN SIDE

NEW GLIMPSES FROM HIS ARCHIVES

Selected and Edited by

Helen Dukas and Banesh Hoffmann

PRINCETON UNIVERSITY PRESS

Published by Princeton University Press, Princeton, New Jersey
In the United Kingdom: Princeton University Press, Guildford, Surrey

Library of Congress Cataloging in Publication Data will be found on the last printed page of this book

This book has been composed in Linotype Primer
Designed by Frank Mahood
Clothbound editions of Princeton University Press books are printed on acid-free paper, and binding materials are chosen for strength and durability.

Printed in the United States of America by Princeton University Press, Princeton, New Jersey

Third printing, with corrections, 1981

First Princeton Paperback printing, 1981

TO OTTO NATHAN

Valued Friend of Albert Einstein
and Dedicated Trustee
of his Estate

Publisher's Preface

PRINCETON University Press plans eventually to publish a complete scholarly edition of *The Writings of Albert Einstein* in about twenty volumes. It will take years for this immense project to be carried out, but preliminary work is under way. In the course of this work we learned that Miss Helen Dukas, Einstein's secretary from 1928 until his death in 1955 and keeper of the Einstein archives since that time, had for her own interest selected various letters and other bits of writing that revealed Einstein's character and personality. Fascinated by the material already collected, we asked Miss Dukas and Dr. Otto Nathan, the trustees of the Estate of Albert Einstein, for permission to publish it and much similar material in a little book. They graciously gave their consent provided that Professor Banesh Hoffmann of Queens College of the City University of New York would agree to collaborate with Miss Dukas on the project. Dr. Hoffmann had been a collaborator of Einstein's, and with Miss Dukas had already written an authoritative biography of Einstein, *Albert Einstein: Creator and Rebel* that won the 1973 Science Writing Award of the American Institute of Physics and the United States Steel Foundation. He and Miss Dukas were delighted to renew their former collaboration, and the result is the present book.

At Princeton University Press we take particular pleasure in this book because we were Einstein's first American publishers, having issued *The Meaning of Relativity* in 1922, and also his last publishers, since the fifth edition of *The Meaning of Relativity* contains Einstein's final formulation of his generalized theory of gravitation. The present book complements the other by giving unique glimpses of a person who was great not only as a scientist but also as a human being.

Albert Einstein
THE HUMAN SIDE

ALBERT EINSTEIN was not only the greatest scientist of his time but also by far the most famous. Moreover, he answered letters. And it is this combination that makes the present book possible.

Unlike our previous book, *Albert Einstein: Creator and Rebel*, this one is not a biography and does not explain Einstein's ideas. It has no chapters, no table of contents, no index, and, at first glance, no plan or structure. It consists, for the most part, of quotations from hitherto unpublished letters and the like that Einstein wrote without thought of publication. There is no need to describe them further here since they speak eloquently for themselves.

Some of the items were sent out in impeccable English, and these we have quoted verbatim. Other items were issued in less idiomatic English, and in presenting them we have made occasional minor changes while preserving the Germanic flavor that gives them charm. All other items are presented in English translation. Often an item that was issued in English was based on a German draft that still exists, and in such cases we have given the English version that was actually sent instead of making an independent translation.

Einstein was an artist not only in his science, which had a transcendent beauty, but also in his use of words. In the latter part of this book, therefore, we have included the original German versions or German drafts, whenever available, so that the reader acquainted with the language can savor Einstein's prose at first hand.

The quest for peace was an important part of Einstein's life. Indeed, a whole book, *Einstein on Peace* (New York, Schocken Books) has been devoted to the subject, and so thorough is its coverage that hardly a scrap of unpublished material on the topic was left over for us to quote. Therefore, for details of this facet of Einstein we refer the reader to that book. We have, however, quoted a lengthy item from the book. Its inclusion has a twofold justification: it is a powerful statement in its own right with a special publication status: and its presence has a symbolic significance as a token salute to all the other items in *Einstein on Peace* that we were sorely tempted to republish here.

The order of presentation of the items is not haphazard. It is akin to that of crowding recollections of a rich life, each sequence apt to take unexpected turns as memory, with a logic all its own, leaps from remembrance to linked remembrance back and forth over the years. In this book there are several such sequences, their starts usually indicated by a more pronounced gap than usual between items. Each item may be taken by itself. But the book is intended to be read as a whole: it offers a seemingly rambling sightseeing journey whose cumulative effect, we hope, will be a deeper and richer understanding of Einstein the man.

For those who would like a road map, we have included a brief Einstein chronology at the end of the book.

O

IN a book about Einstein it is not inappropriate to begin with an item that breaks three rules simultaneously: First, it concerns a letter that Einstein did not answer; second, the presentation makes use of footnotes; and third, the item itself has been published before.

In the summer of 1952 Carl Seelig, an early biographer of Einstein, wrote to him asking for details about his first honorary doctoral degree. In his reply Einstein told of events that occurred in 1909, when Einstein was earning his living at the Swiss patent office in Bern, even though he had propounded his special theory of relativity four years earlier. In the summer of 1909 the University of Geneva bestowed over a hundred honorary degrees in celebration of the 350th anniversary of its founding by Calvin. Here is what Einstein wrote:

> One day I received in the Patent Office in Bern a large envelope out of which there came a sheet of distinguished paper. On it, in picturesque type (I even believe it was in Latin*) was printed something that seemed to me impersonal and of little interest. So right away it went into the official wastepaper basket. Later, I learned that it was an invitation to the Calvin festivities and was also an announcement that I was to receive an honorary doctorate from the Geneva University.** Evidently the people at the university interpreted my silence correctly and

* Actually it was in French, printed in script letters.

** There was a remarkable misprint in the impressive

turned to my friend and student Lucien Chavan, who came from Geneva but was living in Bern. He persuaded me to go to Geneva because it was practically unavoidable—but he did not elaborate further.

So I travelled there on the appointed day and, in the evening in the restaurant of the inn where we were staying, met some Zurich professors. . . . Each of them now told in what capacity he was there. As I remained silent I was asked that question and had to confess that I had not the slightest idea. However, the others knew all about it and let me in on the secret. The next day I was supposed to march in the academic procession. But I had with me only my straw hat and my everyday suit. My proposal that I stay away was categorically rejected, and the festivities turned out to be quite funny so far as my participation was concerned.

The celebration ended with the most opulent banquet that I have ever attended in all my life. So I said to a Genevan patrician who sat next to me, "Do you know what Calvin would have done if he were still here?" When he said no and asked what I thought, I said, "He would have erected a large pyre and had us all burned because of sinful gluttony." The man uttered not another word, and with this ends my recollection of that memorable celebration.

---

document, and this may have registered in Einstein's subconscious and influenced his action: The recipient of the degree was given not as "Monsieur Einstein" but as "Monsieur Tinstein."

In late 1936 the Bern Scientific Society sent Einstein a diploma that had just been awarded to him. On 4 January he wrote back from Princeton as follows:

> You can hardly imagine how delighted I was, and am, that the Bern Scientific Society has so kindly remembered me. It was, as it were, a message from the days of my long-vanished youth. The cosy and stimulating evenings come back to mind once more and especially the often quite wonderful comments that Professor Sahli [Salis?], of internal medicine, used to make about the lectures. I have had the document framed right away and it is the only one of all such tokens of recognition that hangs in my study. It is a memento of my time in Bern and of my friends there.
>
> I ask you to convey my cordial thanks to the members of the Society and to tell them how greatly I appreciate the kindness they have shown me.

A word of amplification is in order: When the document arrived, Einstein said, "This one I want framed and on my wall, because they used to scoff at me and my ideas." Of course, he received numerous other awards. But he did not frame them and hang them on his walls. Instead he hid them away in a corner that he called the "boasting corner" ["Protzenecke"].

In Berlin in 1915, in the midst of World War I, Einstein completed his masterpiece, the general theory of relativity. It not only generalized his special theory of relativity but also provided a new theory of gravitation. Among other things, it predicted the gravitational bending of light rays, which was confirmed by British scientists, notably Arthur Eddington, during an eclipse in 1919. When the confirmation was officially announced, worldwide fame came to Einstein overnight. He never did understand it. That Christmas, writing to his friend Heinrich Zangger in Zurich, he said in part:

> With fame I become more and more stupid, which, of course, is a very common phenomenon. There is far too great a disproportion between what one is and what others think one is, or at least what they say they think one is. But one has to take it all with good humor.

Einstein's fame endured, and brought an extraordinary assortment of mail. For example, a student in Washington, D.C., wrote to him on 3 January 1943 mentioning among other things that she was a little below average in mathematics and had to work at it harder than most of her friends.

Replying in English from Princeton on 7 January 1943, Einstein wrote in part as follows:

> Do not worry about your difficulties in mathematics; I can assure you that mine are still greater.

Back in 1895, after a year away from school, young Einstein became a student at the Swiss Cantonal

School of Argau in the town of Aarau. On 7 November 1896 he sent the following vita to the Argau authorities:

> I was born on 14 March 1879 in Ulm and, when one year old, came to Munich, where I remained till the winter of 1894-95. There I attended the elementary school and the Luitpold secondary school up to but not including the seventh class. Then, till the autumn of last year, I lived in Milan, where I continued my studies on my own. Since last autumn I have attended the Cantonal School in Aarau, and I now take the liberty of presenting myself for the graduation examination. I then plan to study mathematics and physics in the sixth division of the Federal Polytechnic Institute.

Many years later Einstein, now famous, had occasion to prepare another vita. It has some interesting aspects.

Founded in 1652 in the town of Halle was the Kaiser Leopold German Academy of Scientists, in which Goethe had once held membership. On 17 March 1932, in memory of the hundredth anniversary of Goethe's death, the Academy voted to invite Einstein to become a member. When Einstein accepted, the president of the Academy—in accordance with ancient tradition—sent him a biographical questionnaire with nine basic questions. Since space was scarce, Einstein answered in somewhat telegraphic style.

Although the Nazis had not yet come to power, their anti-Semitic propaganda was blatant. Einstein's response to the first question is thus of particular interest. It reads as follows (italics added):

> I. I was born, *the son of Jewish parents*, on 14 March 1879 in Ulm. My father was a merchant, moved shortly after my birth to Munich, in 1893 to Italy, where he remained till his death (1902). I have no brother, but a sister, who lives in Italy.

The second and third questions asked for details of his youth and education, which he dutifully supplied. The fourth question asked about his career. He responded as follows:

> IV. From 1900 to 1902 I was in Switzerland as a private tutor, also for a while employed as house tutor and acquired Swiss citizenship. 1902-09 I was employed as expert (examiner) at the Federal Patent Office, 1909-11 as assistant professor at Zurich University. 1911-12 I was professor of theoretical physics at the Prague University, 1912-14 at the Federal Polytechnic Institute in Zurich also as professor of theoretical physics. Since 1914 I have been a salaried member of the Prussian Academy of Sciences in Berlin and can devote myself exclusively to scientific research work.

The fifth question asked about his achievements and publications. Some of the dates in his answer are puzzling. For example, the special theory of relativity certainly belongs to 1905 and not 1906, and the general theory of relativity to 1915 and not 1916. It is quite possible that Einstein was answering from memory and his memory played tricks on him. Here is what he wrote:

> V. My publications consist almost entirely of short papers in physics, most of which have appeared in the *Annalen der Physik* and the *Proceedings of the Prussian Academy of Sciences*. The most important have to do with the following topics:
>
> Brownian Motion (1905)
> Theory of Planck's formula and Light Quanta (1905, 1917)
> Special Relativity and the Mass of Energy (1906)
> General Relativity 1916 and later.
>
> In addition mention should be made of papers on the thermal fluctuations, as also a [1931] paper, written with Prof. W. Mayer, on the unified nature of gravitation and electricity.

The sixth question asked about scientific travels. He responded as follows:

> VI. Occasional lecture trips to France, Japan, Argentina, England, the United States, which –except for the journeys to Pasadena–did not actually serve research purposes.

The seventh question asked about the goals of his work. He replied:

> VII. The real goal of my research has always been the simplification and unification of the system of theoretical physics. I attained this goal satisfactorily for macroscopic phenomena, but not for the phenomena of quanta and atomic structure. I believe that, despite considerable success, the modern quantum theory is also still far from a satisfactory solution of the latter group of problems.

The eighth question asked about honors that he had received. He answered as follows:

> VIII. I became a member of many, many scientific societies, and several medals were awarded me, also a sort of visiting professorship at the University of Leiden. I have a similar relationship with Oxford University (Christ Church College).

What is extraordinary here is Einstein's failure to mention his 1921 Nobel Prize in Physics. Surely it cannot be ascribed to faulty memory.

The final question was anticlimactic: It asked for his "exact" address.

In school in Aarau, Einstein studied French. Here is a more or less literal translation—after corrections by the French teacher—of an essay that Einstein wrote in that language as an exercise. He was some sixteen years old at the time. The title sounds as if it was assigned by the teacher to the whole class:

My Plans for the Future

A happy man is too contented with the present to think much about the future. But, on the other hand, it is always the young people who like to occupy themselves with bold projects. Besides, it is also a natural thing for a serious young man that he should form for himself as precise an idea as possible of the goal of his desires.

If I have the good fortune to pass my examinations successfully, I shall go to the Federal Institute of Technology in Zurich. I shall stay there four years in order to study mathematics and physics. I imagine myself becoming a professor in those branches of the natural sciences, choosing the theoretical parts of them.

Here are the reasons that have brought me to this plan. Above all is the individual disposition for abstract and mathematical thought, the lack of fantasy and of practical talent. There are also my desires, which have inspired in me the same resolve. This is quite natural. One always likes to do those things for which one has talent. Besides, there is also a certain independence in the scientific profession that greatly pleases me.

In a brief, unpublished biographical essay, Einstein's sister Maja spoke, among other topics, of Einstein's lack of interest in material things of the sort that are often prized by others, and, indeed, almost regarded as necessities. She said, for example: "In his youth he often used to say: 'All I'll want in my dining room is a pine table, a bench, and a few chairs.' "

Here is an excerpt from a letter that Einstein wrote to his sister in 1898, when he was a student in Zurich (he addressed her in his letters as "Dear Sister" much as, later, he would address Queen Elizabeth of Belgium as "Dear Queen"):

> What oppresses me most, of course, is the [financial] misfortune of my poor parents. Also it grieves me deeply that I, a grown man, have to stand idly by, unable to do the least thing to help. I am nothing but a burden to my family. . . . Really, it would have been better if I had never been born. Sometimes the only thought that sustains me and is my only refuge from despair is that I have always done everything I could within my small power, and that year in, year out, I have never permitted myself any amusements or diversions except those afforded by my studies.

Shortly thereafter, in the same year, 1898, with the financial circumstances of his parents somewhat improved, Einstein wrote to his sister as follows:

> There is a fair amount of work to be done, but not too much. So, now and again, I have time to

idle away an hour or so in Zurich's beautiful surroundings. And besides, I am happy in the thought that the worst worries of my parents are now over. If everybody lived as I do, surely the writing of romantic novels would never have come into being. . . .

From the early student days we go to the early days as a member of the Prussian Academy of Sciences in Berlin. In 1918, after the general theory of relativity was completed, the Federal Institute of Technology in Zurich made overtures to see if Einstein would leave Berlin and return to the Polytechnic Institute as a professor there. He wrote to his sister as follows (the dots at the end appear in the original):

> I cannot bring myself to give up everything in Berlin, where people have been so indescribably kind and helpful. How happy I would have been 18 years ago if I had been able to become a lowly assistant at the Federal Institute! But I did not succeed! The world is a madhouse. Renown is everything. After all, other people lecture well too—but . . .

The following letter, also written by Einstein to his sister Maja, belongs to a later time. It is dated 31 August 1935. Much has happened since those early days in Berlin. Einstein is now in Princeton, striving to generalize his general theory of relativity so that it will become a unified field theory. At the same time, all his instincts cry out to him to be wary of developments in the quantum theory that most other physicists accept with equanimity. But his preoccupation with the problems of physics does not blind him to what is going on in the outside world. He writes to his sister as follows:

> As for my work, the going is slow and sticky after a promising start. In the fundamental researches going on in physics we are in a state of groping, nobody having faith in what the other fellow is attempting with high hope. One lives all one's life under constant tension till it is time to go for good. But there remains for me the consolation that the essential part of my work has become part of the accepted basis of our science.
>
> The big political doings of our time are so disheartening that in our generation one feels quite alone. It is as if people had lost the passion for justice and dignity and no longer treasured what better generations have won by extraordinary sacrifices. . . . After all, the foundation of all human values is morality. To have recognized this clearly in primitive times is the unique greatness of our Moses. In contrast, look at the people today! . . .

In 1936 Einstein wrote to his sister as follows:

> I collect nothing but unanswered correspondence and people who, with justice, are dissatis-

fied with me. But can it be otherwise with a man possessed? As in my youth, I sit here endlessly and think and calculate, hoping to unearth deep secrets. The so-called Great World, i.e. men's bustle, has less attraction than ever, so that each day I find myself becoming more of a hermit.

Here are excerpts from a letter that Einstein sent from Berlin to his friend Heinrich Zangger in Zurich in the spring of 1918. The general theory of relativity had already been propounded, but the eclipse verification and world fame were still in the future. Einstein's older son, then some fourteen years of age, was already displaying a lively interest in engineering and technology:

> I, too, was originally supposed to become an engineer. But I found the idea intolerable of having to apply the inventive faculty to matters that make everyday life even more elaborate—and all, just for dreary money-making. Thinking for its own sake, as in music! . . . When I have no special problem to occupy my mind, I love to reconstruct proofs of mathematical and physical theorems that have long been known to me. There is no *goal* in this, merely an opportunity to indulge in the pleasant occupation of thinking. . . .

On 20 August 1949, in answer to a letter asking about his scientific motivation, Einstein wrote, in English:

My scientific work is motivated by an irresistible longing to understand the secrets of nature and by no other feelings. My love for justice and the striving to contribute towards the improvement of human conditions are quite independent from my scientific interests.

Here is a sentence from a letter that Einstein wrote on 13 February 1934 to an interested layman with whom he had corresponded:

As for the search for truth, I know from my own painful searching, with its many blind alleys, how hard it is to take a reliable step, be it ever so small, towards the understanding of that which is truly significant.

In the Berlin days, Einstein often visited Holland, where he had many scientific friends. On a visit to Leyden, Einstein wrote the following in a special memory book for Professor Kammerlingh-Onnes, a pioneer in low-temperature physics who received the Nobel Prize in Physics for 1913. Einstein's note is dated 11 November 1922:

The scientific theorist is not to be envied. For Nature, or more precisely experiment, is an inexorable and not very friendly judge of his work. It never says "Yes" to a theory. In the most favorable cases it says "Maybe," and in the great majority of cases simply "No." If an experiment agrees with a theory it means for the latter "Maybe," and if it does not agree it means "No." Probably every theory will some day experience its "No"– most theories, soon after conception.

Responding to questions put to him by a correspondent in Colorado, Einstein answered as follows on 26 May 1936:

> Outside events capable of determining the direction of a person's thoughts and actions probably occur in everyone's life. But with most people such events have no effect. As for me, when I was a little boy my father showed me a small compass, and the enormous impression that it made on me certainly played a role in my life.
>
> I first learned of the work of Riemann at a time when the basic principles of the general theory of relativity had already long been clearly conceived.

Einstein spoke often of the sense of wonder that came over him when he saw the compass. It was clearly a major event in his life. As for the remark about the work of Riemann, it is of considerable significance. Einstein used Riemann's work as the mathematical basis of the general theory of relativity, and some people have thought that he built on it in the early stages, before the physical concepts had been formulated even in a primitive way. This is, of course, not the only place where Einstein makes the point.

On 17 February 1908, a somewhat aggrieved Einstein in the Patent Office in Bern wrote a postcard to the German physicist Johannes Stark, who was later to receive the Nobel Prize. Here is an excerpt:

> I was somewhat taken aback to see that you did not acknowledge my priority regarding the connection between inertial mass and energy.

This referred to Einstein's now-famous equation $E = mc^2$. On 19 February Stark answered in detail and with a warm display of friendship and admiration, assuring Einstein, the patent examiner, that he spoke favorably of Einstein whenever he could, and that if Einstein thought otherwise he was greatly mistaken. On 22 February 1908 Einstein replied as follows:

> Even if I had not already regretted before receipt of your letter that I had followed the dictates of petty impulse in giving vent to that utterance about priority, your detailed letter really showed me that my over-sensitivity was badly out of place. People who have been privileged to contribute something to the advancement of science should not let such things becloud their joy over the fruits of common endeavor. . . .

Unfortunately, this friendly exchange had a less friendly sequel. With the coming of the Nazis, Stark, like many others, became a bitter doctrinaire critic of Einstein and his works.

In March 1927 Einstein gave a lecture that was taken down verbatim by a member of the audience who suggested to Arnold Berliner, the editor of the scientific journal *Die Naturwissenschaften*, that the lecture be published therein. Berliner consulted Einstein, who responded as follows:

> I am not in favor of its being printed because the lecture is not sufficiently original. One must be especially critical of oneself. One can only continue to expect to be read if, as far as possible, one omits everything that is unimportant.

On 22 February 1949 Einstein wrote the following letter to the writer Max Brod, who was furious because a reviewer had made a mistaken remark about a book of Brod's in the course of a review of Philipp Frank's excellent biography of Einstein:

> Your righteous indignation over the review in *The* [*London*] *Times Literary Supplement* caused me good-natured amusement. Someone, for little pay, and on the basis of a superficial skimming, writes something that sounds halfway plausible and that nobody reads carefully. How can you become serious about such a thing? There have already been published by the bucketful such brazen lies and utter fictions about me that I would long since have gone to my grave if I had let myself pay attention to them. One must console oneself with the thought that Time has a sieve through which most of these important things run into the ocean of oblivion and what remains after this selection is often still trite and bad.

Here is a relevant sentence plucked from a letter that Einstein wrote on 21 March 1930 to his friend Ehrenfest:

> With me, every peep becomes a trumpet solo.

And here is one from a letter to his biographer Carl Seelig, written on 25 October 1953:

> In the past it never occurred to me that every casual remark of mine would be snatched up and recorded. Otherwise I would have crept further into my shell.

Einstein found some aspects of the English quite puzzling. For instance, Helen Dukas, Einstein's secretary, vividly recalls that back in 1930, during a short stopover of their ship at Southampton on the way to the United States, a British reporter asked her if he could have an interview with Einstein. Aware of Einstein's wishes, she said "No" and braced herself for battle. But, to her surprise, he accepted the "No" without argument and left. This was not an isolated incident. Other British reporters behaved similarly on the same occasion. She mentioned this to Einstein, and it prompted part of the following entry in his travel diary:

> 3 December 1930 (Southampton): . . . In England even the reporters are reserved! Honor to whom honor is due. A single "No" is sufficient.

> The world can still learn a lot here—except that I don't want it, and always dress sloppily, even at the holy sacrament of dinner.

Later, Professor F. A. Lindemann, who was to become a scientific adviser to Winston Churchill, arranged for Einstein to visit Oxford. Einstein stayed at Christ Church College, where the rituals were not markedly different from those in other Oxford colleges. Like most of the others, Christ Church was for men only. The rooms were chilly. Each evening the dons and students—five hundred of them—in academic gowns, gathered formally in the great hall for dinner, with grace read out in Latin. Here is part of an entry in Einstein's travel diary:

> Oxford, 2/3 May 1931: Calm existence in [my] cell while freezing badly. Evening: solemn dinner of the holy brotherhood in tails.

Here is an entry of a different sort, telling of a storm at sea:

> 10 December 1931: Never before have I lived through a storm like the one this night. . . . The sea has a look of indescribable grandeur, especially when the sun falls on it.
>
> One feels as if one is dissolved and merged into Nature. Even more than usual, one feels the insignificance of the individual, and it makes one happy.

In sending an etching of himself to Dr. Hans Mühsam, a medical friend in Berlin, Einstein wrote the following underneath the portrait. The date is 1920 or perhaps earlier, and the etching was made by Hermann Struck:

> Measured objectively, what a man can wrest from Truth by passionate striving is utterly infinitesimal. But the striving frees us from the bonds of the self and makes us comrades of those who are the best and the greatest.

To his friend Paul Ehrenfest, who like himself was a theoretical physicist, Einstein wrote the following sentence in a letter dated 15 March 1922:

> How wretchedly inadequate is the theoretical physicist as he stands before Nature—and before his students!

o

IN Princeton, early in December 1950, Einstein received a long handwritten letter from a nineteen-year-old student at Rutgers University who said "My problem is this, sir, 'What is the purpose of man on earth?' " Dismissing such possible answers as to make money, to achieve fame, and to help others, the student said "Frankly, sir, I don't even know why I'm going to college and studying engineering." He felt that man is here "for no purpose at all" and went on to quote from Blaise Pascal's *Pensées* the following words, which he said aptly summed up his own feelings: "I know not who put me into the world, nor what the world is, nor what I myself am. I am in terrible ignorance of everything. I know not what my body is, nor my senses, nor my soul, not even that part of me which thinks what I say, which reflects on all and on itself, and knows itself no more than the rest. I see those frightful spaces of the universe which surround me, and I find myself tied to one corner of this vast expanse, without knowing why I am put in this place rather than another, nor why this short time which is given me to live is assigned to me at this point rather than at another of the whole eternity which was before me or which shall come after me. I see nothing but infinities on all sides, which surround me as an atom, and as a shadow which endures only for an instant and returns no more. All I know is that I must die, but what I know least is this very death which I cannot escape."

The student remarked that Pascal saw the answers to be in religion but that he himself did not. After elaborating on the cosmic insignificance of man, he nevertheless asked Einstein to tell him where the right course lay, and why, saying "Pull no punches. If you think I've gone off the track let me have it straight."

In responding to this poignant cry for help, Einstein offered no easy solace, and this very fact must have heartened the student and lightened the lonely burden of his doubts. Here is Einstein's response. It was written in English and sent from Princeton on 3 December 1950, within days of receiving the letter:

> I was impressed by the earnestness of your struggle to find a purpose for the life of the individual and of mankind as a whole. In my opinion there can be no reasonable answer if the question is put this way. If we speak of the purpose and goal of an action we mean simply the question: which kind of desire should we fulfill by the action or its consequences or which undesired consequences should be prevented? We can, of course, also speak in a clear way of the goal of an action from the standpoint of a community to which the individual belongs. In such cases the goal of the action has also to do at least indirectly with fulfillment of desires of the individuals which constitute a society.
>
> If you ask for the purpose or goal of society as a whole or of an individual taken as a whole the question loses its meaning. This is, of course, even more so if you ask the purpose or meaning of nature in general. For in those cases it seems quite arbitrary if not unreasonable to assume

somebody whose desires are connected with the happenings.

Nevertheless we all feel that it is indeed very reasonable and important to ask ourselves how we should try to conduct our lives. The answer is, in my opinion: satisfaction of the desires and needs of all, as far as this can be achieved, and achievement of harmony and beauty in the human relationships. This presupposes a good deal of conscious thought and of self-education. It is undeniable that the enlightened Greeks and the old Oriental sages had achieved a higher level in this all-important field than what is alive in our schools and universities.

On 28 October 1951 a graduate student in psychology sent a beautifully worded letter to Einstein in Princeton asking for advice. The student was an only child and, like his parents, Jewish though not orthodox. A year and a half before, he had fallen deeply in love with a girl of the Baptist faith. Knowing the pitfalls in a mixed marriage, and the unintended wounds that could be inflicted by the thoughtless remarks of others, the couple had mixed socially with friends and acquaintances and found that their love was able to withstand the stresses. The girl, unprompted, had expressed a willingness to convert to Judaism so that their children would have a more homogeneous family life. While the young man's parents liked the girl, they were frightened of intermarriage and gave voice to their objections. The young man was torn between his love for the girl and his desire not to alienate his parents and cause them lasting pain. He asked whether he was not right in believing that a wife takes precedence over parents when one ventures upon a new mode of life.

Einstein drafted a reply in German on the back of the letter. The reply may very well have been sent in English, but only the German draft is in the Einstein Archives. Here is a translation of it:

> I have to tell you frankly that I do not approve of parents exerting influence on decisions of their children that will determine the shapes of the children's lives. Such problems one must solve for oneself.
>
> However, if you want to make a decision with which your parents are not in accord, you must ask yourself this question: Am I, deep down, independent enough to be able to act against the wishes of my parents without losing my inner equilibrium? If you do not feel certain about this, the step you plan is also not to be recommended in the interests of the girl. On this alone should your decision depend.

On 8 December 1952 a twenty-year-old student majoring in philosophy at Brown University sent Einstein, in Princeton, a long, enthusiastic handwritten letter telling eloquently that he had been a profound admirer of Einstein since as far back as he could remember, and that all matters concerning Einstein—his theories, his views, his personality—had long had an overwhelming fascination for him. He wondered if

Einstein could possibly find time to write him a brief note. Since Einstein did not know him, the student realized that it could not contain a personal message, but he hoped it would contain a message or statement of some sort nevertheless.

On 9 December 1952 Einstein replied in English as follows:

> It is the most beautiful reward for one who has striven his whole life to grasp some little bit of truth if he sees that other men have real understanding of and pleasure with his work. I thank you therefore very much for your kind words. Having little time to spare I must be content to write you only a short remark.
>
> It is true that the grasping of truth is not possible without empirical basis. However, the deeper we penetrate and the more extensive and embracing our theories become the less empirical knowledge is needed to determine those theories.

On 4 October 1931 Einstein gave a lecture at the Berlin Planetarium. A correspondent, unable to attend, read about the lecture in the newspaper the next day and sent him the clipping. Here is its account of what Einstein said:

> For the creation of a theory the mere collection of recorded phenomena never suffices—there must always be added a free invention of the human mind that attacks the heart of the matter. And: the physicist must not be content with the purely

phenomenological considerations that pertain to the phenomena. Instead, he should press on to the speculative method, which looks for the underlying pattern.

The Einsteins had a summer home in Caputh, near Berlin, which they greatly enjoyed. It was later to be confiscated by the Nazis, and even in 1932 the future looked dark. The daughter of a neighbor in Caputh had an album to which she asked Einstein to contribute. He did so in 1932 with these words:

> O Youth: Do you know that yours is not the first generation to yearn for a life full of beauty and freedom? Do you know that all your ancestors felt as you do—and fell victim to trouble and hatred?
>
> Do you know, also, that your fervent wishes can only find fulfillment if you succeed in attaining love and understanding of men, and animals, and plants, and stars, so that every joy becomes your joy and every pain your pain? Open your eyes, your heart, your hands, and avoid the poison your forebears so greedily sucked in from History. Then will all the earth be your fatherland, and all your work and effort spread forth blessings.

A fifth grade teacher in Ohio found that most of his students were shocked to learn that human beings are classed as belonging to the animal kingdom. He persuaded them to compose letters asking the opinions of great minds and, on 26 November 1952, he sent a selection to Einstein in Princeton hoping that Einstein would find time to reply. This Einstein did, in English, on 17 January 1953, as follows:

Dear Children,

We should not ask "What is an animal" but "What sort of a thing do we call an animal?" Well, we call something an animal which has certain characteristics: it takes nourishment, it descends from parents similar to itself, it grows, it moves by itself, it dies if its time has run out. That's why we call the worm, the chicken, the dog, the monkey an animal. What about us humans? Think about it in the above-mentioned way and then decide for yourselves whether it is a natural thing to regard ourselves as animals.

On 25 February 1952, representatives of the "Sixth Form Society" of a grammar school in England wrote to Einstein telling enthusiastically that he had been elected almost unanimously to the Rectorship of their group. Admittedly, the office involved no duties, and, indeed, according to the bylaws the group was not allowed to have a Rector anyway. But they felt that Einstein would appreciate the gesture as an indication of their recognition of the greatness of his work.

On 17 March 1952 Einstein replied in English as follows:

As an old schoolmaster I received with great joy and pride the nomination to the Office of

> Rectorship of your society. Despite my being an old gypsy there is a tendency to respectability inherent in old age—so with me. I have to tell you, though, that I am a little (but not too much) bewildered by the fact that this nomination was made independent of my consent.

Einstein's letter was framed and placed in the school Library, where the "Sixth Form Society" used to meet. It is probably there still.

A child in the sixth grade in a Sunday School in New York City, with the encouragement of her teacher, wrote to Einstein in Princeton on 19 January 1936 asking him whether scientists pray, and if so what they pray for. Einstein replied as follows on 24 January 1936:

> I have tried to respond to your question as simply as I could. Here is my answer.
>
> Scientific research is based on the idea that everything that takes place is determined by laws of nature, and therefore this holds for the actions of people. For this reason, a research scientist will hardly be inclined to believe that events could be influenced by a prayer, i.e. by a wish addressed to a supernatural Being.
>
> However, it must be admitted that our actual knowledge of these laws is only imperfect and fragmentary, so that, actually, the belief in the existence of basic all-embracing laws in Nature also rests on a sort of faith. All the same this

faith has been largely justified so far by the success of scientific research.

But, on the other hand, every one who is seriously involved in the pursuit of science becomes convinced that a spirit is manifest in the laws of the Universe—a spirit vastly superior to that of man, and one in the face of which we with our modest powers must feel humble. In this way the pursuit of science leads to a religious feeling of a special sort, which is indeed quite different from the religiosity of someone more naive.

It is worth mentioning that this letter was written a decade after the advent of Heisenberg's principle of indeterminacy and the probabilistic interpretation of quantum mechanics with its denial of strict determinism.

The following letter, sent by Einstein from Princeton on 20 December 1935, is self-explanatory. This is fortunate because there seems to be no record of the circumstances that gave rise to it. The letter may well have been the result of an oral request:

Dear Children,

It gives me great pleasure to picture you children joined together in joyful festivities in the radiance of Christmas lights. Think also of the teachings of him whose birth you celebrate by these festivities. Those teachings are so simple —and yet in almost 2000 years they have failed to prevail among men. Learn to be happy through the happiness and joy of your fellows, and not through the dreary conflict of man against man! If you can find room within yourselves for this natural feeling, your every burden in life will be light, or at least bearable, and you will find your way in patience and without fear, and will spread joy everywhere.

In answer to an oral question from a child, transmitted by her mother, Einstein wrote the following in English on 19 June 1951:

> There has been an earth for a little more than a billion years. As for the question of the end of it I advise: Wait and see!

He added a postscript saying:

> I enclose a few stamps for your collection.

○

IN Dresden a government official, who spoke of himself as a politician and an Adlerian psychotherapist, was planning a book to be based on psychoanalyses of important people. In this connection, on 17 January 1927, he wrote to Einstein in Berlin to ask if he would allow himself to be psychoanalyzed.

It is not known if an answer was actually sent, but, on the letter, in German in Einstein's handwriting, is the following draft of a reply:

> I regret that I cannot accede to your request, because I should like very much to remain in the darkness of not having been analyzed.

At first, Einstein was not favorably inclined toward the work of Sigmund Freud, though later he changed his mind. On Einstein's fiftieth birthday, Freud, like many others, sent him greetings. In his note he spoke of Einstein as "you lucky one" ["Sie Glücklicher"], a phrase that aroused Einstein's curiosity. Einstein in Berlin replied on 22 March 1929 as follows:

> Revered Master,
>
> I thank you warmly for having thought of me. Why do you stress my "luck"? You, who have slipped under the skin of so many a man—indeed, of man*kind*—have nevertheless had no opportunity to slip under mine.
>
> With highest regard and cordial good wishes,

In reply, Freud explained that he regarded Einstein as lucky because nobody unfamiliar with physics would dare to judge his work, whereas Freud's own work was subject to judgment by everybody, whether familiar with psychology or not.

O

ON 20 January 1921 the editor of a German magazine dealing with modern art wrote to Einstein in Berlin saying that he himself was convinced that there was a close connection between the artistic developments and the scientific results belonging to a given epoch. He asked Einstein to write a few paragraphs on the subject for publication in his magazine. On 27 January 1921 Einstein replied in these words:

> Although I am aware that I have nothing original, let alone worthy of publication, to say on the theme you mention, I send you the enclosed aphoristic utterance to demonstrate my good will. If my ink had been less viscid, I would have done justice to the wish expressed in your friendly letter by sending a more sumptuous opus.

The "aphoristic utterance," which was published in the magazine, went as follows:

What Artistic and Scientific Experience Have in Common

> Where the world ceases to be the scene of our personal hopes and wishes, where we face it as free beings admiring, asking, and observing, there we enter the realm of Art and Science. If what is seen and experienced is portrayed in the language of logic, we are engaged in science. If it is communicated through forms whose connections are not accessible to the conscious mind but are

recognized intuitively as meaningful, then we are engaged in art. Common to both is the loving devotion to that which transcends personal concerns and volition.

An afterword: When the Nazis came to power, the editor, who was not a Jew, tried to flee Germany. On being stopped at the border, he killed himself.

The following two aphorisms were jotted down by Einstein in Huntington, N.Y., in 1937. While almost certainly not inspired by the preceding item, they are not unrelated to it:

> Body and soul are not two different things, but only two different ways of perceiving the same thing. Similarly, physics and psychology are only different attempts to link our experiences together by way of systematic thought.

> Politics is a pendulum whose swings between anarchy and tyranny are fueled by perennially rejuvenated illusions.

The following aphorism, in English, was attributed to Einstein by a South American writer who used it as a motto at the head of his letter. Since it faithfully echoes frequent conversational remarks made by Einstein, it may be accepted as genuine. Only the English version was given by the writer:

> Nationalism is an infantile sickness. It is the measles of the human race.

On 17 July 1953 a woman who was a licensed Baptist pastor sent Einstein in Princeton a warmly appreciative evangelical letter. Quoting several passages from the scriptures, she asked him whether he had considered the relationship of his immortal soul to its Creator, and asked whether he felt assurance of everlasting life with God after death. It is not known whether a reply was sent, but the letter is in the Einstein Archives, and on it, in Einstein's handwriting, is the following sentence, written in English:

> I do not believe in immortality of the individual, and I consider ethics to be an exclusively human concern with no superhuman authority behind it.

In 1954 or 1955 Einstein received a letter citing a statement of his and a seemingly contradictory statement by a noted evolutionist concerning the place of intelligence in the Universe. Here is a translation of the German draft of a reply. It is not known whether a reply was actually sent:

> The misunderstanding here is due to a faulty translation of a German text, in particular the use of the word "mystical." I have never imputed to Nature a purpose or a goal, or anything that could be understood as anthropomorphic.
>
> What I see in Nature is a magnificent structure that we can comprehend only very imperfectly, and that must fill a thinking person with a feeling of "humility." This is a genuinely religious feeling that has nothing to do with mysticism.

In Berlin in February 1921 Einstein received from a woman in Vienna a letter imploring him to tell her if he had formed an opinion as to whether the soul exists and with it personal, individual development after death. There were other questions of a similar sort. On 5 February 1921 Einstein answered at some length. Here in part is what he said:

> The mystical trend of our time, which shows itself particularly in the rampant growth of the so-called Theosophy and Spiritualism, is for me no more than a symptom of weakness and confusion.
>
> Since our inner experiences consist of reproductions and combinations of sensory impressions, the concept of a soul without a body seems to me to be empty and devoid of meaning.

An official of a subsidiary of the American publishing house McGraw-Hill was due to address an annual conference of the American Library Association. On 1 April 1948 he wrote to Einstein for help, saying that librarians and publishers alike had become alarmed at a widespread decline in interest in books on science for the *layman*. He asked Einstein to express an opinion as to the reasons for the decline, and mentioned that a similar letter had been sent to other outstanding scientists and science writers. Einstein, who had long held strong views about the popularization of science, lost no time in replying. On 3 April he sent the following in English:

> The situation, in my opinion, is as follows:

> Most books about science that are said to be written for the layman seek more to impress the reader ("awe-inspiring!" "how far we have progressed!" etc.), than to explain to him clearly and lucidly the elementary aims and methods. After an intelligent layman has tried to read a couple of such books he becomes completely discouraged. His conclusion is: I am too feebleminded and had better give up. In addition, the entire description is done mostly in a sensational manner which also repulses a sensible layman.
>
> In one word: Not the readers are at fault but the authors and the publishers. My proposition is: no "popular" book on science should be published before it is established that it can be understood and appreciated by an intelligent and judicious layman.

The above letter seems not to have been published before. There may be some point in quoting here the start of a letter that Einstein sent in English to *Popular Science Monthly* on 28 January 1952. It was published in that journal. The editor had had a letter from an awe-struck reader asking about work of Einstein's concerning which Einstein had said, according to the reader, that it would "solve the secrets of the universe." The editor asked Einstein to answer the reader's questions, which he did in simple, undramatic terms. But he could not resist making the following remarks at the start of his letter:

> It is not my fault that laymen obtain an exaggerated impression of the significance of my efforts. Rather, this is due to writers of popular

science and in particular to newspaper correspondents who present everything as sensationally as possible.

Here are two items that we present as a pair. Inevitably, Einstein received an enormous number of letters from people who believed that they had an idea of major scientific importance. Sometimes his patience ran out. This is a case in point. On 7 July 1952 a letter was written to him by an artist in New York City. On 10 July 1952 Einstein sent the following reply in English from Princeton:

> Thank you for your letter of July 7th. You seem to be a living container of all the empty expressions in vogue among the intellectuals of this country. If I would be a dictator I would forbid the use of all these unfortunate inanities.

On 22 March 1954 a self-made man sent Einstein in Princeton a long handwritten letter–four closely packed pages in English. The correspondent despaired that there were so few people like Einstein who had the courage to speak out, and he wondered if it would not be best to return the world to the animals. Saying "I presume you would like to know who I am," he went on to tell in detail how he had come from Italy to the United States at the age of nine, arriving in bitter cold weather, as a result of which his sisters died while he barely survived; how after six months of schooling he went to work at age ten; how at age seventeen he went to Evening School; and so on, so that now he had a regular job as an experimental machinist, had a spare-time business of his own, and had some patents to his credit. He declared himself an atheist. He said that real education came from reading books. He cited an article about Einstein's religious beliefs and expressed doubts as to the article's accuracy. He was irreverent about various aspects of formal religion, speaking about the millions of people who prayed to God in many languages, and remarking that God must have an enormous clerical staff to keep track of all their sins. And he ended with a long discussion of the social and political systems of Italy and the United States that it would take too long to describe here. He also enclosed a check for Einstein to give to charity.

On 24 March 1954 Einstein answered in English as follows:

> I get hundreds and hundreds of letters but seldom one so interesting as yours. I believe that your opinions about our society are quite reasonable.
>
> It was, of course, a lie what you read about my religious convictions, a lie which is being systematically repeated. I do not believe in a personal God and I have never denied this but have expressed it clearly. If something is in me which can be called religious then it is the unbounded admiration for the structure of the world so far as our science can reveal it.
>
> I have no possibility to bring the money you sent me to the appropriate receiver. I return

it therefore in recognition of your good heart and intention.

Your letter shows me also that wisdom is not a product of schooling but of the lifelong attempt to acquire it.

In September 1920, Einstein visited Stuttgart to give a lecture. While there, his wife Elsa invited all their cousins for a visit and a drive but unfortunately left out the cousins' young children, one of these being eight-year-old Elisabeth Ney. Knowing that the young girl had a sense of humor, Einstein sent her, on 30 September 1920, the following bantering post-card. She cherished it, which is why it still exists:

Dear Miss Ney,

I hear from Elsa that you are dissatisfied because you did not see your uncle Einstein. Let me therefore tell you what I look like: pale face, long hair, and a tiny beginning of a paunch. In addition an awkward gait, and a cigar in the mouth—if he happens to have a cigar—and a pen in his pocket or his hand. But crooked legs and warts he does not have, and so he is quite handsome—also no hair on his hands such as is often found on ugly men. So it is indeed a pity that you did not see me.

With warm greetings from
Your Uncle Einstein.

On 12 April 1950 a distant relative of Einstein's wrote to him from Paris to say that the writer's son was about to enter the university to study physics and chemistry and was eager to receive a few words from the most famous member of the family.

On 18 May 1950 Einstein replied as follows, beginning with a little rhyme:

> The way things stand, my embarrassment's
> distressing.
> Were I a parson, I'd gladly give my blessing.

> However, I am glad to have heard from you, and also to have learned that your son wants to devote himself to the study of physics. But I cannot refrain from telling you that it is a difficult matter if one is not going to be satisfied with superficial results. It is best, it seems to me, to separate one's inner striving from one's trade as far as possible. It is not good when one's daily bread is tied to God's special blessing.

The years passed, and on 1 March 1954 the relative wrote once more to Einstein to tell what had been happening in the meantime. The son had had Einstein's letter framed and had hung it over the bed in his study. Einstein's words, said the relative, had clearly had magical power because the son had passed his first diploma examination at the head of his class. When offered a reward of his choice, such as a ski vacation or money, the son had timidly wondered if he could perhaps have a signed photograph of his famous protector and ideal.

The photograph, signed, was duly sent.

On 11 July 1947 an Idaho farmer wrote to Einstein telling that he had given his son the name Albert and wondering if Einstein would write a few words that he could keep "as a talisman" to encourage the son as he grew up. Einstein responded in English on 30 July 1947 with these words:

> Nothing truly valuable arises from ambition or from a mere sense of duty; it stems rather from love and devotion towards men and towards objective things.

Enclosing a snapshot of little Albert, the delighted father wrote back saying that, as a token of appreciation, he was sending Einstein a sack of Idaho potatoes. It turned out to be quite a big sack.

o

PARTLY because of a history-making series of scientific meetings in Brussels, partly because of a shared love of music, and above all because of mutual regard, a remarkable friendship sprang up between Einstein and King Albert and Queen Elizabeth of Belgium. The nature of the friendship is shown vividly in the following excerpts from a letter written by Einstein in Brussels in 1931 to his wife Elsa. In the letter Einstein tells of a visit to the palace.

> I was received with touching warmth. These two people are of a purity and kindness seldom found. First we talked for about an hour. Then [the Queen and I] played quartets and trios [with an English lady musician and a musical lady-in-waiting]. This went on merrily for several hours. Then they all went away and I stayed behind alone for dinner at the King's—vegetarian style, no servants. Spinach and fried eggs and potatoes, period. (It had not been anticipated that I would stay.) I liked it very much there, and I am certain the feeling is mutual.

The friendship with the royal family of Belgium endured and deepened. In a letter to Einstein in Germany dated 30 July 1932, Queen Elizabeth enclosed copies of photographs that she had taken of him, told him how much she had enjoyed talking with him and wandering in the park, and said she had not forgotten his lucid explanation of the causal and probabilistic theories in physics. On 19 September 1932 Einstein replied in part as follows:

> It gave me great pleasure to tell you about the mysteries with which physics confronts us. As a human being, one has been endowed with just enough intelligence to be able to see clearly how utterly inadequate that intelligence is when confronted with what exists. If such humility could be conveyed to everybody, the world of human activities would be more appealing.

On 9 February 1931 Einstein wrote to the Queen from Santa Barbara, California, as follows:

> Two days I have been staying in this carefree corner where wind and heat and cold are all unknown. And yesterday I was shown a dreamy villa (Bliss) in which you are said to have spent a few happy, quiet days some years ago.
>
> It is now two months that I have been in this country of contradictions and surprises, where one alternates between admiration and head-shaking. One realizes that one is attached to the old Europe with its problems and its pains, and returns gladly.

Two years later, on 19 February 1933, Einstein, again in Santa Barbara, sent the Queen a small twig together with a quatrain of which the following is a translation:

> In cloister garden a small tree stands.
> Planted by your very hands.

It sends—its greetings to convey—
A twig, for it itself must stay.

The Queen replied in kind from the Palace in Laeken on 15 March 1933. By that time the Nazis had come to power and had confiscated Einstein's money and reviled him widely. Near the end of her rhyme the Queen obliquely refers to this and also plays on the name Einstein, which, when written as two words "Ein Stein," means "one stone." Here is a translation of what she wrote:

"The twig the greeting did convey
From the tree that had to stay,
And from the friend, whose heart so big
Could send great joy by tiny twig.
A thousand thanks aloud I cry
Unto mountain, sea, and sky.
Now, when stones begin to shake,
I pray ONE STONE no harm will take."

In January 1934 the Einsteins, now safely settled in Princeton, were the guests of President and Mrs. Roosevelt in the White House. In the course of the conversation there were cordial remembrances of the Queen. Einstein wanted her to know of this, so he wrote the following verse hailing the President. It was dated 25 January 1934, and was sent to the Queen on White House stationery. Below, with minor change, is the official translation of the rhyme:

In the Capital's proud glory
Where Destiny unfolds her story,

Fights a man with happy pride
Who solutions can provide.

In our talk of yester night
Memories of you were bright;
In remembrance of our meeting
Let me send you this rhymed greeting.

When the Einsteins returned to Europe from Pasadena in 1933, shortly after the Nazis seized power in Germany, it was not wholly by chance that they took refuge in the tiny Atlantic resort of Le Coq-sur-mer. For Le Coq was in Belgium. King Albert and Queen Elizabeth were deeply concerned about Einstein's safety. With rumors flying that the Nazis had placed a price on Einstein's head, King Albert ordered that there be two bodyguards to protect him day and night.

The following letter from Einstein in Princeton to Elizabeth comes a little later and offers a glimpse of yet another facet of the friendship. The letter makes brief mention of "the Barjanskys." Mr. and Mrs. Barjansky, friends of the Einsteins, were also friends of the Belgian royal couple. Indeed, Mr. Barjansky played cello in the Queen's quartet, and his wife, a sculptor, gave lessons to the Queen. It was through their intervention that the letter came to be written. These are the circumstances: In the spring of 1934 King Albert fell to his death while mountain climbing, and in late summer the following year the new Queen, Astrid—Elizabeth's daughter-in-law—died in an automobile accident at the age of thirty. The double blow devastated Elizabeth. She fell into a state of profound emotional numbness, unable even to bring herself to make music with the quartet or to work on her sculpture, to the alarm of those around her.

Mrs. Barjansky wrote to Einstein telling him of the situation and suggesting that a letter from him to Elizabeth might do some good. Here is Einstein's letter. It is dated 20 March, and although the year is not given it was almost certainly 1936:

> Dear Queen,
>
> Today, for the first time this year, the spring sunshine has made its appearance, and it aroused me from the dreamlike trance into which people like myself fall when immersed in scientific work. Thoughts rise up from an earlier and more colorful life, and with them comes remembrance of beautiful hours in Brussels.
>
> Mrs. Barjansky wrote to me how gravely living in itself causes you suffering and how numbed you are by the indescribably painful blows that have befallen you.
>
> And yet we should not grieve for those who have gone from us in the primes of their lives after happy and fruitful years of activity, and who have been privileged to accomplish in full measure their task in life.
>
> Something there is that can refresh and revivify older people: joy in the activities of the younger generation—a joy, to be sure, that is clouded by dark forebodings in these unsettled times. And yet, as always, the springtime sun brings forth

new life, and we may rejoice because of this new life and contribute to its unfolding; and Mozart remains as beautiful and tender as he always was and always will be. There is, after all, something eternal that lies beyond reach of the hand of fate and of all human delusions. And such eternals lie closer to an older person than to a younger one oscillating between fear and hope. For us, there remains the privilege of experiencing beauty and truth in their purest forms.

Have you ever read the Maxims of La Rochefoucauld? They seem quite acerb and gloomy, but by their objectivization of human and all-too-human nature they bring a strange feeling of liberation. In La Rochefoucauld we see a man who succeeded in liberating himself even though it had not been easy for him to be rid of the heavy burden of the passions that Nature had dealt him for his passage through life. It would be nicest to read him with people whose little boat has gone through many storms: for example, the good Barjanskys. I would gladly join in were it not forbidden by "the big water."

I am privileged by fate to live here in Princeton as if on an island that in many respects resembles the charming palace garden in Laeken. Into this small university town, too, the chaotic voices of human strife barely penetrate. I am almost ashamed to be living in such peace while all the rest struggle and suffer. But after all, it is still the best to concern oneself with eternals, for from them alone flows that spirit that can restore peace and serenity to the world of humans.

With my heartfelt hope that the spring will bring quiet joy to you also, and will stimulate you to activity, I send you my best wishes.

O

LITTLE is known about the following item. From its content one can reasonably assume that it belongs to the spring or summer of 1933, when Einstein was in Le Coq. There is about it an air of the ludicrous since Einstein was hardly the man to contemplate resort to physical violence:

> You ask me what I thought when I heard that the Potsdam police had invaded my summer home to search for hidden weapons.
>
> What else would a Nazi policeman assume?

(While the above more or less conveys the meaning, it is far from being close to the original. A literal translation of the last sentence of the German would read: "I am reminded of the German proverb: Everyone measures according to his own shoes." The point of the proverb, of course, is that we evaluate others according to our own lights—that we expect them to be like ourselves.)

On one occasion Einstein remarked that a quiet and lonely job like that of a lighthouse keeper would be ideal for a reflective scholar or a theoretical physicist. For an Einstein this could well be so. But what of his apparent assumption that others would also flourish under such austere circumstances? It may well remind us of a German proverb.

Here are two items that, each in its own way, must have lightened Einstein's spirits in the dark days of the Nazi takeover.

Having heard that Einstein's property in Germany had been confiscated by the Nazis, the Dutch astronomer W. de Sitter, in the name of his colleagues, wrote to Einstein offering financial assistance. On 5 April 1933 Einstein replied as follows:

> In times like these one has an opportunity to learn who are one's true friends. I thank you warmly for your readiness to help. But actually things are going very well with me so that not only can I manage for myself and mine with what I have but also I can help others keep their heads above water. From Germany, however, I will probably not be able to rescue anything because an action is being taken against me for high treason. The physiologist [Jacques] Loeb once said to me in conversation that political leaders must all really be pathological because a normal person would not be able to bear so tremendous a responsibility while being so little able to foresee the consequences of his decisions and acts. Although this may have sounded somewhat exaggerated at the time, it turns out to be true in full measure of Germany today. The only curious thing is the utter failure of the so-called intellectual aristocracy [in Germany].

During a visit to England in 1933, after having left Germany for good and just before going to the United States to take up his position at the Institute for Advanced Study in Princeton, N.J., Einstein received a letter from a correspondent whose knowledge of physics could hardly be characterized as sound.

For example, the correspondent said that, according to his understanding, the world moves so fast that it seems to be stationary. He went on to say, in all seriousness, that because of gravity a person on the spherical earth is sometimes upright, sometimes standing on his head, sometimes sticking out at right angles to the earth, and sometimes, as he put it, "at left angles." And he asked if perhaps it was while upside down, standing on their heads, that people fell in love and did other foolish things.

So far as is known, the letter was not answered. But on it, in German, Einstein did jot down the following words:

> Falling in love is not at all the most stupid thing that people do—but gravitation cannot be held responsible for it.

In Princeton, shortly after his arrival there, Einstein was asked by a freshman publication, *The Dink*, for a message. He responded with these words, which were published in December 1933:

> I am delighted to live among you young and happy people. If an old student may say a few words to you they would be these: Never regard your study as a duty, but as the enviable opportunity to learn to know the liberating influence of beauty in the realm of the spirit for your own personal joy and to the profit of the community to which your later work belongs.

On 24 March 1951 a student at a college in California wrote to Einstein in Princeton wondering if he recalled dedicating the little observatory there. She went on to ask him for advice. She had long had a deep interest in astronomy and wanted to become a professional astronomer. But two of her teachers had told her that there were already too many astronomers, and that anyway she was not good enough to be able to succeed in the field. Conceding that she was not outstanding in mathematics, she asked Einstein whether she should go on with astronomy or seek something else she might pursue.

Einstein replied in English as follows:

> Science is a wonderful thing if one does not have to earn one's living at it. One should earn one's living by work of which one is sure one is capable. Only when we do not have to be accountable to anybody can we find joy in scientific endeavor.

While this advice may seem to have been specially constructed for a student of whom Einstein knew little, he regarded it as fundamental and broadly applicable. He well knew the strain of being expected to produce new ideas. When invited to Berlin he likened himself to a hen that is expected to continue to lay eggs. He urged often that the would-be scientist or scholar earn his living in a non-demanding job like that of a cobbler, thus avoiding the "publish or perish" pressures that undermined joy in one's creative work and led one to publish superficial results.

After all, the great philosopher Spinoza, whom he revered, had earned his living as a lens grinder, and Einstein himself often recalled with nostalgic pleasure the days when he was earning his living by working at the patent office in Bern while producing some of his greatest ideas.

The following item offers further illustration of this theme.

On 14 July 1953 an Indian in Delhi wrote Einstein a long and somewhat repetitious letter seeking aid. The gist of the letter was this: The writer was a bachelor, 32 years old, who wanted to spend the rest of his life doing nothing but research in mathematics and physics, although he was, admittedly, "terribly weak" in those subjects. He was penniless, as witness the fact that he was not placing stamps on the envelope of his letter. Financial stringency in his young days had prevented him from acquiring a good foundation in science and mathematics although his interest in them was ardent. Family circumstances had obliged him to take a job to earn a living—something that was strongly against his inner nature. Fortunately, because of a small quarrel, he had been dismissed from his job more than a year before, and was thus free to pursue his true mission—but, alas, with no income at all to keep body and soul together. With or without help, he was determined to continue till he died, but, of course, life would be easier with financial aid, and he hoped that Einstein would help him.

On 28 July 1953 Einstein replied to him in English at some length in a letter that is interesting for more than just its courtesy:

I received your letter and was impressed by your ardent wish to continue to study physics. I must confess, however, that I can in no way agree with your attitude. We all are nourished and housed by the work of our fellow-men and we have to pay honestly for it not only by work chosen for the sake of our inner satisfaction but by work which, according to general opinion, serves them. Otherwise one becomes a parasite however modest our wants might be. This is the more so in your country where the work of educated persons is doubly needed in this time of struggle for economic improvement.

This is one side of the matter. But there is another side to it which would have to be considered also in the case that you would have ample means to choose freely what to do. In striving to do scientific work the chance—even for very gifted persons—to achieve something of real value is very little, so that it would always be a great probability that you would feel frustrated when the age of optimal working capacity has passed.

There is only one way out: Give most of your time to some practical work as a teacher or in another field which agrees with your nature, and spend the rest of it for study. So you will be able, in any case, to lead a normal and harmonious life even without the special blessings of the Muses.

Einstein's dislike of the academic pressure to produce extended to the rat race for promotion. On 5 May 1927, at a time when the scientific world was wondering who would succeed to the professorship held by Planck at the University of Berlin, Einstein wrote this to his friend Paul Ehrenfest in Holland:

> I am not involved, thank God, and no longer need to take part in the competition of the big brains. Participating in it has always seemed to me to be an awful type of slavery no less evil than the passion for money or power.

A handwritten statement by Einstein was found in the Central Zionist Archives in Jerusalem. It was written on 3 October 1933, the day on which he and other distinguished persons spoke at a mass meeting in the Albert Hall in London to obtain aid for scholars who were refugees from Nazi Germany. Shortly thereafter, Einstein left England for the United States, never to return to Europe. It is not known to whom the message was addressed. In it one sees a reflection of the mass dismissals of Jewish scholars by the Nazis:

> The value of Judaism lies exclusively in its spiritual and ethical content and in the corresponding qualities of individual Jews. For this reason, from olden times till now, study has rightly been the sacred endeavor of the capable ones among us. That is not to say, however, that we should earn our livelihood by intellectual work, as is, unfortunately, too often the case among us. In these dire times we must do our utmost to adjust to practical necessity without giving up love of the

spiritual and intellectual and cultivation of our studies.

In a letter to Einstein in Princeton dated 30 March 1935 a correspondent quoted the following, which the *New York Herald Tribune* had attributed to Einstein: "There are no German Jews; there are no Russian Jews; there are no American Jews; there are only Jews." He felt that Einstein must have been misquoted, pointing out, for example, that "people of the Jewish creed" fought in German, Russian, and American armies and staunchly upheld the causes of their respective nations.

On 3 April 1935 Einstein replied as follows:

> In the last analysis, everyone is a human being, irrespective of whether he is an American or a German, a Jew or a Gentile. If it were possible to manage with this point of view, which is the only dignified one, I would be a happy man. I find it very sad that divisions according to citizenship and cultural tradition should play so great a role in modern practical life. But since this cannot be changed, one should not close one's eyes to reality.
>
> Now as to the Jews and their ancient traditional community: seen through the eyes of the historian, their history of suffering teaches us that the fact of being a Jew has had a greater impact than the fact of belonging to political communities. If, for example, the German Jews were driven from Germany, they would cease to be Germans and would change their language and their political

> affiliation; but they would remain Jews. Why this is so is certainly a difficult question. I see the reason not so much in racial characteristics as in firmly rooted traditions that are by no means limited to the area of religion. This state of affairs is not changed by the fact that Jews, as citizens of specific states, lose their lives in the wars of those states.

In the above letter there is no specific mention of Zionism. And yet Zionism was already much in Einstein's mind. Early in 1919—before the eclipse verification of the general theory of relativity, and thus before Einstein's world fame—Kurt Blumenfeld, a Zionist official, had broached the subject to him. Two years later Blumenfeld persuaded Einstein to accept an invitation from Chaim Weizmann to travel to the United States with Weizmann in order to raise funds for the creation of a Hebrew University in Jerusalem. Weizmann, the world leader of the Zionist movement —he was later to become the first president of the State of Israel—was himself a scientist. Telling about the boat trip across the Atlantic he said, "Einstein explained his theory to me every day, and on my arrival I was fully convinced that he understood it."

Here is part of a letter that Einstein wrote to his friend Heinrich Zangger on 14 March 1921:

> On Saturday I'm off to America—not to speak at universities (though there will probably be that, too, on the side) but rather to help in the founding of the Jewish University in Jerusalem. I feel an intense need to do something for this cause.

And here is part of a letter he sent to his physicist friend Paul Ehrenfest on 18 June 1921:

> Zionism indeed represents a new Jewish ideal that can restore to the Jewish people their joy in existence. . . . I am very happy that I accepted Weizmann's invitation.

Einstein had become a figure of enormous symbolic importance to Jews. In 1923, when he visited Mount Scopus, the site on which the Hebrew University was to rise, he was invited to speak from "the lectern that has waited for you for two thousand years."

In a letter to Paul Ehrenfest dated 12 April 1926 Einstein wrote concerning the Hebrew University:

> I do believe that in time this endeavor will grow into something splendid; and, Jewish Saint that I am, my heart rejoices.

Responding to a Jewish anti-Zionist, probably in January 1946, Einstein wrote:

> In my opinion, condemning the Zionist movement as "nationalistic" is unjustified. Consider the path by which Theodor Herzl came to his mission. Initially he had been completely cosmopolitan. But during the Dreyfus trial in Paris he suddenly realized with great clarity how precarious was the situation of the Jews in the western world. And courageously he drew the conclusion that we

are discriminated against or murdered not because we are Germans, Frenchmen, Americans, etc. of the "Jewish faith" but simply because we are Jews. Thus already our precarious situation forces us to stand together irrespective of our citizenship. Zionism gave the German Jews no great protection against annihilation. But it did give the survivors the inner strength to endure the debacle with dignity and without losing their healthy self-respect. Keep in mind that perhaps a similar fate could be lying in wait for your children.

And in March of 1955, less than a month before he died, Einstein wrote these words to Kurt Blumenfeld, mentioned above, who had introduced him to Zionism:

> I thank you, even at this late hour, for having helped me become aware of my Jewish soul.

Sam Gronemann was a man of many parts: a Berlin lawyer, an author, a playwright, and a prominent Zionist who left Nazi Germany to settle in Israel. On 13 March 1949, for Einstein's seventieth birthday, he sent a letter from Tel Aviv in Israel containing verses of which the following is a translation:

"One who, after struggling many a night,
Can't get Relativity quite right;
Yes, one who has no feeling of alarm,
Because coordinates never did him harm,

He argues thus, and does so most politely:
If seventy-year-old Einstein still feels sprightly,
Then one can irrefutably deduce
That Einstein's theory has a practical use.
With friendly throngs surrounding you to pay
Homage on your seventieth natal day,
I, too, will show no doubts or hesitations
In offering *us* and you congratulations.
For we in Israel, as you'll understand,
Think of you as *ours* in this our land."

Einstein responded forthwith. Here is a translation of his response:

Non-comprehenders are often distressed.
Not you, though—because with good humor
you're blessed.
After all, your thought went like this, I dare
say:
It was none but the Lord who made
us that way.

The Lord takes revenge—and it's simply unfair,
For he himself made the weakness we bear.
And lacking defense we succumb to this
badness,
Sometimes in triumph, and sometimes in
sadness.

But rather than stubbornly uttering curses,
You bring us salvation by means of your verses,
Which are cunningly made so the just and
the sinners
End up by counting themselves all as winners.

There is in the Einstein Archives a letter dated 5 August 1927 from a banker in Colorado to Einstein in Berlin. Since it begins "Several months ago I wrote you as follows," one may assume that Einstein had not yet answered. The banker remarked that most scientists and the like had given up the idea of God as a bearded, benevolent father figure surrounded by angels, although many sincere people worship and revere such a God. The question of God had arisen in the course of a discussion in a literary group, and some of the members decided to ask eminent men to send their views in a form that would be suitable for publication. He added that some twenty-four Nobel Prize winners had already responded, and he hoped that Einstein would too. On the letter, Einstein wrote the following in German. It may or may not have been sent:

> I cannot conceive of a personal God who would directly influence the actions of individuals, or would directly sit in judgment on creatures of his own creation. I cannot do this in spite of the fact that mechanistic causality has, to a certain extent, been placed in doubt by modern science.
>
> My religiosity consists in a humble admiration of the infinitely superior spirit that reveals itself in the little that we, with our weak and transitory understanding, can comprehend of reality. Morality is of the highest importance—but for us, not for God.

Here is an excerpt from a letter that Einstein wrote to Cornelius Lanczos on 24 January 1938. As will be seen, it is highly relevant to the present topic:

> I began with a skeptical empiricism more or less like that of Mach. But the problem of gravitation converted me into a believing rationalist, that is, into someone who searches for the only reliable source of Truth in mathematical simplicity.

In speaking of gravitation, Einstein is here referring to his general theory of relativity, the fruit of ten years of inspired labor, from 1905 to 1915. It came about because of a feeling of aesthetic unease. According to the special theory of relativity of 1905, uniform motion was relative. Einstein felt that it was ugly to have only one special type of motion relative. If uniform motion was relative, then *all* motion should be. But common, everyday experience showed that non-uniform motion was absolute. In the face of such evidence a lesser man would have shrugged his shoulders and decided that there was nothing to do but tolerate his aesthetic discomfort. But not Einstein. Driven by an aesthetic compulsion, he looked anew at the everyday evidence and saw, to his surprise and delight, that it could be interpreted as showing that all motion could indeed be regarded as relative. This is not the place to tell how this insight led him to gravitational equations of transcendental beauty. But we can begin to see what Einstein had in mind when, in his letter to Lanczos, he said that he had been converted into a believing rationalist—a seeker after mathematical simplicity, by which he meant beauty.

Let us not be confused by the word "converted." Einstein was seeking beauty in the Universe long before he created his general theory of relativity; this is clear from the fact that that theory arose from an aesthetic discontent. His belief—his religious belief—in the simplicity, beauty, and sublimity of the Universe was the primary source of inspiration in his science. He evaluated a scientific theory by asking himself whether, if he were God, he would have made the Universe in that way.

Here are excerpts from two other letters from Einstein in Princeton to Lanczos. On 14 February 1938 he wrote:

> I have now struggled with this basic problem of electricity for more than twenty years, and have become quite discouraged, though without being able to let go of it. I am convinced that a completely new and enlightening inspiration is needed; I also believe, on the other hand, that the flight into statistics is to be regarded only as a temporary expedient that by-passes the fundamentals.

On 21 March 1942 he wrote:

> You are the only person I know who has the same attitude towards physics as I have: belief in the comprehension of reality through something basically simple and unified. . . . It seems hard to sneak a look at God's cards. But that he plays dice and uses "telepathic" methods (as the present quantum theory requires of him) is something that I cannot believe for a single moment.

We see here Einstein's vivid way of looking at and expressing his dissatisfaction with the quantum theory, with its denial of determinism and its limitation to probabilistic, statistical predictions. He was himself a pioneer in the development of the quantum theory, but he remained convinced that there was need for a different understanding. He vividly expressed his frustration, which never left him, in this excerpt from a letter that he wrote to Paul Ehrenfest on 12 July 1924:

> The more one chases after quanta, the better they hide themselves.

A Chicago Rabbi, preparing a lecture on "The Religious Implications of the Theory of Relativity," wrote to Einstein in Princeton on 20 December 1939 to ask some questions on the topic. Einstein replied as follows:

> I do not believe that the basic ideas of the theory of relativity can lay claim to a relationship with the religious sphere that is different from that of scientific knowledge in general. I see this connection in the fact that profound interrelationships in the objective world can be comprehended through simple logical concepts. To be sure, in the theory of relativity this is the case in particularly full measure.
>
> The religious feeling engendered by experiencing the logical comprehensibility of profound interrelations is of a somewhat different sort from the

feeling that one usually calls religious. It is more a feeling of awe at the scheme that is manifested in the material universe. It does not lead us to take the step of fashioning a god-like being in our own image—a personage who makes demands of us and who takes an interest in us as individuals. There is in this neither a will nor a goal, nor a must, but only sheer being. For this reason, people of our type see in morality a purely human matter, albeit the most important in the human sphere.

The following statement by Einstein is dated September 1937. Beyond the fact that it has to do with a "Preaching Mission," nothing of any consequence is known of the circumstances that prompted its composition. It may have been written in response to a personal request from a member of the Princeton Theological Seminary, but that is pure conjecture:

Our time is distinguished by wonderful achievements in the fields of scientific understanding and the technical application of those insights. Who would not be cheered by this? But let us not forget that knowledge and skills alone cannot lead humanity to a happy and dignified life. Humanity has every reason to place the proclaimers of high moral standards and values above the discoverers of objective truth. What humanity owes to personalities like Buddha, Moses, and Jesus ranks for me higher than all the achievements of the enquiring and constructive mind.

What these blessed men have given us we must

guard and try to keep alive with all our strength if humanity is not to lose its dignity, the security of its existence, and its joy in living.

The German draft of the following was among many items brought to Le Coq from Einstein's last stay in Pasadena in the winter of 1932-33. No date appears on the slip of paper, and no indication of the occasion. The words may have been in response to a letter from an individual or a group, or they may have been an aphorism prompted by a Nazi boast. Certainly, they stand well on their own as a message addressed to us all:

> Do not pride yourself on the few great men who, over the centuries, have been born on your earth—through no merit of yours. Reflect, rather, on how you treated them at the time, and how you have followed their teachings.

On 25 February 1931 a correspondent wrote a deeply pessimistic letter to Einstein in Berlin telling of disillusion with the technological marvels of the time, declaring that for most people life was a bitter disappointment, and wondering whether it was sensible to propagate mankind. On 7 April 1931 Einstein replied as follows:

> I do not share your opinion. I have always felt that my own life was interesting and worth living, and I am firmly convinced that it is both possible

and likely that people's lives in general can be made worth living. The objective and psychological possibilities for this seem to be there.

Of course, Einstein was aware that sorrow is a part of life. On 26 April 1945 he wrote the following letter of condolence to a medical doctor and his wife who had lost a grandchild, or perhaps a child. The doctor had been active in helping refugees from Nazi Germany:

> I am profoundly shocked by the news of the terrible blow that has so suddenly and unexpectedly befallen you both. This is the most grievous thing that can happen to older people, and it is no consolation that untold thousands are similarly afflicted. I would not dare presume to try to comfort you, but I do want to tell you how deeply and with what sorrow I sympathize with you, as do all those who have come to know the kindness of your heart.
>
> For the most part we humans live with the false impression of security and a feeling of being at home in a seemingly familiar and trustworthy physical and human environment. But when the expected course of everyday life is interrupted, we realize that we are like shipwrecked people trying to keep their balance on a miserable plank in the open sea, having forgotten where they came from and not knowing whither they are drifting. But once we fully accept this, life becomes easier and there is no longer any disappointment.
>
> Here's hoping that the planks on which we are swimming will meet again soon, Cordially,

Here is a sentence from a letter that Einstein sent to Cornelius Lanczos on 9 July 1952:

> One is born into a herd of buffaloes and must be glad if one is not trampled underfoot before one's time.

The botanist A. V. Fric found a small, hitherto unknown flowering cactus plant in the rarefied atmosphere on the highest mountain peak of the Cordilleras. In a graciously worded report, he gave it the name "Einsteinia" and sent Einstein a copy of the report. On 9 September 1933 Einstein replied as follows from Le Coq:

> Dear Sir,
>
> You have given me great pleasure by your thoughtful act. The naming is apt to the extent that not just the little plant but I too have not been left in peace at the aethereal summit.
>
> In grateful appreciation of your gratifying gesture, I am,

Here is a translation of lines that were inscribed and autographed by Einstein below a photograph of himself that he sent to an old friend, Mrs. Cornelia Wolf, in 1927:

> Wherever I go and wherever I stay,
> There's always a picture of me on display.
> On top of the desk, or out in the hall,
> Tied round a neck, or hung on the wall.

Women and men, they play a strange game,
Asking, beseeching: "Please sign your name."
From the erudite fellow they brook not a quibble,
But firmly insist on a piece of his scribble.

Sometimes, surrounded by all this good cheer,
I'm puzzled by some of the things that I hear,
And wonder, my mind for a moment not hazy,
If I and not they could really be crazy.

There is an interesting story connected with this inscribed photograph. During World War II, Mrs. Wolf crossed the ocean on her way to Havana and subsequently to California. The boat stopped in Trinidad, where a British officer interrogated her (she had a German passport) and began to examine her baggage. Although she knew that the British did not permit passengers to take along any photographs or letters, she had not been able to bring herself to leave the Einstein photograph behind. When the officer found it, he at once stopped interrogating her and asked very politely whether he could borrow it, to copy the poem and to show his colleagues. She told him that it was in his power even to keep it, but he replied that he would return it the next day before the ship sailed. This he did, with the utmost courtesy, and there was no further interrogation or search of baggage.

o

EINSTEIN was an ardent violinist, his violin being his inseparable companion. The composers of the eighteenth century were his favorites. He loved the music of Bach and Mozart. Beethoven he admired rather than loved, and he felt less rapport with later composers.

With fame came an intense and often annoying interest in all aspects of his life. So when a German illustrated weekly in 1928 sent him a questionnaire in Berlin about Johann Sebastian Bach it is not surprising that Einstein ignored it. The editor waited a while and then, on 24 March 1928, sent a second request. This time, on the very same day—the mails were faster in those days—Einstein answered brusquely as follows:

> This is what I have to say about Bach's life work: listen, play, love, revere—and keep your mouth shut.

It so happened that later that same year a different publication asked Einstein about a different composer, and on 10 November 1928 Einstein replied as follows:

> As to Schubert, I have only this to say: play the music, love—and shut your mouth!

Some ten years later a more searching questionnaire about Einstein's musical tastes arrived from a yet different source and this one he answered in more detail. The questionnaire seems to have been lost, but the questions on it can be inferred more or less from Einstein's responses, which bear as date only the year 1939:

> (1) Bach, Mozart, and some old Italian and English composers are my favorites in music. Beethoven considerably less—but certainly Schubert.
>
> (2) It is impossible for me to say whether Bach or Mozart means more to me. In music I do not look for logic. I am quite intuitive on the whole and know no theories. I never like a work if I cannot intuitively grasp its inner unity (architecture).
>
> (3) I always feel that Handel is good—even perfect—but that he has a certain shallowness. Beethoven is for me too dramatic and too personal.
>
> (4) Schubert is one of my favorites because of his superlative ability to express emotion and his enormous powers of melodic invention. But in his larger works I am disturbed by a certain lack of architectonics.
>
> (5) Schumann is attractive to me in his smaller works because of their originality and richness of feeling, but his lack of formal greatness prevents my full enjoyment. In Mendelssohn I perceive considerable talent but an indefinable lack of depth that often leads to banality.
>
> (6) I find a few lieder and chamber works by Brahms truly significant, also in their structure. But most of his works have for me no inner

persuasiveness. I do not understand why it was necessary to write them.

(7) I admire Wagner's inventiveness, but I see his lack of architectural structure as decadence. Moreover, to me his musical personality is indescribably offensive so that for the most part I can listen to him only with disgust.

(8) I feel that [Richard] Strauss is gifted, but without inner truth and concerned only with outside effects. I cannot say that I care nothing for modern music in general. I feel that Debussy is delicately colorful but shows a poverty of structure. I cannot work up great enthusiasm for something of that sort.

For the most part, as we see, the works of the modern composers of his day held little appeal for Einstein. However, he had high personal regard for Ernst Bloch, and on 15 November 1950, apparently responding to a request for a statement, he wrote this in English:

My knowledge of modern music is very restricted. But in one respect I feel certain: True art is characterized by an irresistible urge in the creative artist. I can feel this urge in Ernst Bloch's work as in few later musicians.

When the great conductor Arturo Toscanini was presented with the American Hebrew Medal in January 1938, Einstein wrote the following statement, which was presumably read aloud at the time of the presentation:

> Only one who devotes himself to a cause with his whole strength and soul can be a true master. For this reason mastery demands all of a person. Toscanini demonstrates this in every manifestation of his life.

In October 1928 a correspondent wrote to Einstein in Berlin asking him among other things whether his musical activity had any influence on his main field of work, which was so different. On 23 October 1928 Einstein replied as follows:

> Music does not *influence* research work, but both are nourished by the same source of longing, and they complement one another in the release they offer.

o

DR. Otto Juliusburger, a friend of Einstein, was by profession a psychiatrist with a practice in Berlin. He was also something of an expert on Spinoza and Schopenhauer. As a Jew he was aware of increasing danger. In 1937 he was able to send his two children, one at a time, to the United States, and almost at the last moment before the advent of the infamous gas chambers, the parents succeeded in following their children. Below are excerpts, on various topics, from letters from Einstein to Juliusburger, and one from a letter from Juliusburger to Einstein.

On 28 September 1937 Einstein wrote from Princeton to Juliusburger, who was still in Berlin, telling of his pleasure that the son had already arrived in the United States and saying that he had heard encouraging words as to the possible admission of the daughter. After speaking of other matters, he ended by talking about his current research endeavors, which had to do with his long search for a unified field theory that would link gravitation and electromagnetism. Here are the closing paragraphs of the letter:

> I still struggle with the same problems as ten years ago. I succeed in small matters but the real goal remains unattainable, even though it sometimes seems palpably close. It is hard and yet rewarding: hard because the goal is beyond my powers, but rewarding because it makes one immune to the distractions of everyday life.
>
> I can no longer accommodate myself to the

people here and their way of life. I was already too old to do so when I came over, and to tell the truth it was no different in Berlin, and before that in Switzerland. One is born a loner, as you will understand, being one yourself.

Here is a letter that Einstein wrote to Juliusburger on 2 August 1941. The Juliusburgers were now safely in the United States:

> I count myself fortunate to be able to greet you here after all these years. I had imposed silence on myself because any note from me to a person in Barbaria would have exposed him to danger. Your beloved Schopenhauer once said that people in their misery are unable to achieve tragedy but are condemned to remain stuck in tragi-comedy. How true it is, and how often I have felt this impression. Yesterday idolized, today hated and spit upon, tomorrow forgotten, and the day after tomorrow promoted to Sainthood. The only salvation is a sense of humor, and we will keep that as long as we still draw breath.

On 30 September 1942 Einstein wrote to Juliusburger as follows, the greetings mentioned in the first paragraph probably being for the Jewish New Year:

> I was greatly moved by your kind words and I send you belated greetings. I know that I do not in the least deserve so much praise, but I enjoyed the cordial feelings that glow from your words.

I believe that we can now at last hope to see the day when the unspeakable wrong may be somewhat expiated. But all the misery, all the desperation, all the senseless annihilations of human lives—nothing can make up for that. And yet we may hope that now even the dullest creature can be made to realize that, in the long run, lies and tyranny cannot triumph.

One sees in you the unshakable fortitude that the search for truth can bestow. I, too, owe to such an attitude my only true satisfactions. One feels that in the timeless community of people of this sort one finds a sort of refuge that fends off desperation and feelings of hopeless isolation.

In a letter to Einstein in Princeton written in September 1942, Juliusburger spoke of the funeral services, some fifteen years earlier, for Einstein's mother-in-law, and recalled that as they were leaving the grave Einstein had said to him:

The closing words of the beautiful prayer, "The Lord gave, and the Lord hath taken away; blessed be the name of the Lord" signify the abundance of life that always gives and always takes back—in order to give again.

On 11 April 1946 Einstein wrote to Juliusburger as follows:

You take a definite stand about Hitler's responsibility. I myself have never really believed in the subtler distinctions that lawyers foist upon physicians. Objectively, there is, after all, no free will.

> I think that we have to safeguard ourselves against people who are a menace to others, quite apart from what may have motivated their deeds. What need is there for a criterion of responsibility? I believe that the horrifying deterioration in the ethical conduct of people today stems primarily from the mechanization and dehumanization of our lives—a disastrous byproduct of the development of the scientific and technical mentality. Nostra culpa! I don't see any way to tackle this disastrous short-coming. Man grows cold faster than the planet he inhabits.

And on 29 September 1947 Einstein wrote to Juliusburger as follows:

> I hear from several friends that these days you are celebrating—I can hardly believe it possible!—your eightieth birthday. People like you and me, though mortal, of course, like everyone else, do not grow old no matter how long we live. What I mean is that we never cease to stand like curious children before the great Mystery into which we are born. This interposes a distance between us and all that is unsatisfactory in the human sphere—and this is no small matter. When, in the mornings, I become nauseated by the news that the *New York Times* sets before us, I always reflect that it is anyway better than the Hitlerism that we only barely managed to finish off.

The letter of 29 September 1947 that Einstein wrote to Juliusburger is reminiscent of a much earlier item. Professor Federigo Enriques had arranged a scientific congress in Bologna that Einstein attended, and there he met the professor's daughter, Adriana. She may have asked him for an autographed note. Whether she did or not, he wrote her the following signed handwritten message in October of 1921:

> Study and in general the pursuit of truth and beauty is a sphere of activity in which we are permitted to remain children all our lives.
>
> To Adrianna Enriques, a memento of our acquaintanceship of October 1921.

Einstein's letter of 11 April 1946 to Juliusburger ties in with the second of the following two letters that deal with the death penalty.

In a letter to a publisher in Berlin that Einstein wrote on 3 November 1927 with regard to an earlier statement on the subject, he said:

> I have reached the conviction that the abolition of the death penalty is desirable.
> Reasons:
>
> (1) Irreparability in the event of an error of justice,
>
> (2) Detrimental moral influence of the execution procedure on those who, whether directly or indirectly, have to do with the procedure.

Einstein returned to the subject in a letter dated 4 November 1931 in answer to a letter from a troubled young man in Prague. Here is an excerpt:

You ask me what I think about war and the death penalty. The latter question is simpler. I am not for punishment at all, but only for measures that serve society and its protection. In principle I would not be opposed to killing individuals who are worthless or dangerous in that sense. I am against it only because I do not trust people, i.e. the courts. What I value in life is quality rather than quantity, just as in Nature the overall principles represent a higher reality than does the single object.

On 1 February 1954 a correspondent cited Einstein's urgings that people be prepared to go to prison if necessary in order to preserve free speech and to oppose war. The correspondent went on to say that his wife, seeing what Einstein had written, pointed out that he had not wasted any time in leaving Germany when the Nazis came to power instead of staying on to speak out and risk being jailed. She contrasted this with the behavior of Socrates, who refused to leave his country but stayed to "fight it out." She also remarked that it was easier for well-known people to speak out than it was for lesser men.

On 6 February 1954 Einstein replied in English as follows (For some reason he omitted in the English version a remark in the German draft, a translation of which is added here enclosed in square brackets.):

Thank you for your letter of February 1st. I think what your wife has said is pretty well to the point. It is true that a man who enjoys some popularity is in a less precarious situation than someone unknown to the public. But what better use could a person make of his "name" than to speak out publicly from time to time if he believes it necessary?

The comparison with Sokrates is not quite to the point. For Sokrates Athens meant the World. I, however, never identified myself with any particular country, least of all with the German state with which my only connection was my position as member of the Prussion Academy of Sciences [and the language, which I learned as a child].

Although I am a convinced democrat I know well that the human community would stagnate and even degenerate without a minority of socially conscious and upright men and women willing to make sacrifices for their convictions. Under present circumstances this holds true to a higher degree than in normal times. You will understand this without explanation.

Einstein held Supreme Court Justice Louis D. Brandeis in the highest esteem. Here is part of a brief statement that he sent from Caputh on 19 October 1931 to the Boston journal *The Jewish Advocate*, which was celebrating Brandeis's seventy-fifth birthday:

True human progress is based less on the inventive mind than on the conscience of men such as Brandeis.

On 10 November 1936 Einstein, in Princeton, sent him directly the following letter (the handwritten original is among the Brandeis papers at the Law School of the University of Louisville):

> With deepest veneration and fellow feeling, I clasp your hand on the occasion of your eightieth birthday. I know of no other person who combines such profound intellectual gifts with such self-renunciation while finding the whole meaning of his life in quiet service to the community. We—all of us—thank you not only for what you have accomplished and brought about, but also because we feel happy that such a man should exist at all in this time of ours, which is so lacking in genuine personalities.
>
> With reverent greetings. . . .

Walter White, who was secretary of the National Association for the Advancement of Colored People, was white not only in name but also in skin color. Had he wished to pass as a white he could certainly have done so, and thereby avoided all the troubles and persecutions that then even more than now were the lot of the Negro in our society. But he chose instead to fight for the rights of his black brethren, knowing full well what the price would be in personal suffering. In 1947 he wrote a moving article entitled "Why I remain a Negro" that was published in *The Saturday Review of Literature* in the issue of 11 October. It prompted the following comment from Einstein to the editors. Here is the official translation:

On reading the White article one is struck with the deep meaning of the saying: There is only one road to true human greatness—the road through suffering. If the suffering springs from the blindness and dullness of a tradition-bound society, it usually degrades the weak to a state of blind hate, but exalts the strong to a moral superiority and magnanimity which would otherwise be almost beyond the reach of man.

I believe that every sensitive reader will, as I did myself, put down Walter White's article with a feeling of true thankfulness. He has allowed us to accompany him on the painful road to human greatness by giving us a simple biographical story which is irresistible in its convincing power.

The following letter was sent in English by Einstein on 4 November 1942 to a correspondent in Brazil. It is self-explanatory:

Your proposition seems to me reasonable in principle: organization of the economy through a small number of people who have proved themselves as productive and as intensely and unselfishly interested in the improvement of the prevailing conditions. I do not believe, however, in your method of selection by "tests." This is a typical engineer-idea which does not correspond with your own statement that "man is not a machine."

Furthermore, I want you to consider one thing: It is not enough to find the ten best suited persons—one must also get the nations to submit to their decisions and decrees. How this should be achieved I have no idea. This problem is far more difficult than the choice of suitable personalities. Even rather mediocre people would be able to achieve the goal in a passable way compared with the conditions as they exist at present and have existed up to now. Hitherto the leaders came to their power mainly not by their ability to think and to make decisions but by their faculty to impress, to persuade and to use the shortcomings of their fellow-beings.

The old problem: what should be done to give the power into the hands of capable and well-meaning persons has so far resisted all efforts. Unfortunately as far as I can see you too have not found a way to solve it.

On 6 December 1917, during World War I, Einstein wrote the following from Berlin to Heinrich Zangger in Zurich. The words have not become dated:

How is it at all possible that this culture-loving era could be so monstrously amoral? More and more I come to value charity and love of one's fellow being above everything else. . . . All our lauded technological progress—our very civilization—is like the axe in the hand of the pathological criminal.

In 1934 Einstein wrote an article on the subject of *Tolerance* for a magazine in America. When the editors wanted to make changes that were not to his liking, he withdrew the article and it was not published. Here are excerpts from it:

> As I now ask myself what tolerance really is, there comes to mind the amusing definition that the humorous Wilhelm Busch gave of "abstinence":
>
> > Abstinence is the pleasure we net
> > From various things we do not get.
>
> I could say analogously that tolerance is the affable appreciation of qualities, views, and actions of other individuals which are foreign to one's own habits, beliefs, and tastes. Thus being tolerant does not mean being indifferent towards the actions and feelings of others. Understanding and empathy must also be present. . . .
>
> Whether it be a work of art or a significant scientific achievement, that which is great and noble comes from the solitary personality. European culture made its most important break away from stifling stagnation when the Renaissance offered the individual the possibility of unfettered development.
>
> The most important kind of tolerance, therefore, is tolerance of the individual by society and the state. The state is certainly necessary, in order to give the individual the security he needs for his development. But when the state becomes the main thing and the individual becomes its weak-willed tool, then all finer values are lost. Just as the rock must first crumble for trees to grow on it, and just as the soil must first be loosened for its fruit-

fulness to develop, so too can valuable achievement sprout from human society only when it is sufficiently loosened so as to make possible to the individual the free development of his abilities.

Sometimes Einstein's own tolerance was sorely strained. In the following we see him seeking relief in biting satire.

It is extraordinary that something so highly technical and abstruse as Einstein's theory of relativity should become the target of political attacks. The attacks were often virulent. In Germany, the Nazis condemned the theory as being Jewish and Communistic, and said that it poisoned the well-springs of German science. And, of course, they forbade scientists to teach it. Only a few brave souls dared to defy the order, and even they resorted to stratagems such as presenting the ideas without mentioning Einstein or using the word "relativity."

As for the Soviet Union, that country was not at all as sure as the Nazis were that Einstein's theory was Communistic. Indeed, the official Russian attitude toward the theory was complicated by debate as to whether it was in accord with Dialectical Materialism, the philosophical basis of Marxism. As a result, it was not always safe for a Soviet scientist to espouse the theory. The situation is better now, but even as late as April 1952 a member of the USSR Academy of Sciences charged that Einstein had dragged physics into "the swamp of idealism," that Einstein was guilty of "subjectivism," and that, in contrast, Dialectical Materialism was based on "the objectivity of material nature." Moreover, the Academician publicly criticized by name two Russian scientists whom he accused of favoring the theory. The attack was reported widely by the Associated Press, and a longtime friend in London sent Einstein an account of it that had appeared in Berlin.

The following unpublished satirical comment was found among Einstein's papers. It is believed to belong to the early 1950's and was almost certainly inspired by the Soviet attitude in general and by the above incident in particular.

> When Almighty God was laying down His eternal laws of Nature, He became troubled by the following doubt, which He was unable to overcome even afterwards: How awkward a situation would result if, later on, the High Authorities of Dialectical Materialism were to declare some, or even all of His laws unlawful.
>
> Later, when He went on to create the Prophets and Wise Men of Dialectical Materialism, a somewhat analogous doubt sneaked into His soul. He quickly regained His composure, however, when He realized that He could be confident that those Prophets and Wise Men would never come to the conclusion that the tenets of Dialectical Materialism could be in contradiction with Reason and Truth.

A correspondent in England wrote to Einstein in Germany asking him a question that had originally been posed by Edison. It asked: If, on your death bed, you looked back on your life, by what facts would you determine whether it was a success or failure? On 12 November 1930, Einstein replied as follows:

> Neither on my death bed nor before will I ask myself such a question. Nature is not an engineer or contractor, and I myself am a part of Nature.

o

ON 13 November 1950, the minister of a church in Brooklyn, N.Y., wrote to Einstein in Princeton saying, among other things, that some twenty-six years before, as a college student, he had bought an autographed photograph of Einstein that he had cherished ever since. He went on to say that shortly after the rise of Hitler, Einstein had made a statement that the minister had often quoted from his pulpit. He wondered if Einstein would send him the two paragraphs of the statement copied out in Einstein's own handwriting so that he could reframe the photograph to include them.

Saying that he did not wish to seem like a parasite, he enclosed a check–not in payment, since he knew that one could not buy such a handwritten statement, but as a gift for Einstein to use as he saw fit, and as a token of gratitude. On a separate sheet he copied out the statement he was referring to. Here it is:

"Being a lover of freedom, when the revolution came to Germany I looked to the universities to defend it, knowing that they had always boasted of their devotion to the cause of truth; but no, the universities were immediately silenced. Then I looked to the great editors of the newspapers whose flaming editorials in those days gone by had proclaimed their love of freedom; but they, like the universities, were silenced in a few short weeks. Then I looked to the individual writers who, as literary guides of Germany, had written much and often concerning the place of freedom in modern life; but they, too, were mute.

"Only the Church stood squarely across the path

of Hitler's campaign for suppressing the truth. I never had any special interest in the Church before, but now I feel a great affection and admiration because the Church alone has had the courage and persistence to stand for intellectual truth and moral freedom. I am forced to confess that what I once despised I now praise unreservedly."

On 14 November 1950, Einstein replied in English as follows:

> I was deeply impressed with the fine and generous way you have approached me in your letter of November 11th. I am, however, a little embarrassed. The wording of the statement you have quoted is not my own. Shortly after Hitler came to power in Germany I had an oral conversation with a newspaper man about these matters. Since then my remarks have been elaborated and exaggerated nearly beyond recognition. I cannot in good conscience write down the statement you sent me as my own.
>
> The matter is all the more embarrassing to me because I, like yourself, am predominantly critical concerning the activities, and especially the political activities, through history of the official clergy. Thus, my former statement, even if reduced to my actual words (which I do not remember in detail) gives a wrong impression of my general attitude.
>
> I am, however, gladly willing to write something else which would suit your purpose, if you give me any indication what it could be.

The minister replied on 16 November 1950 saying he was glad the statement had not been correct since he too had had reservations about the historical role of the Church at large. He elaborated at some length on the topic and then suddenly apologized for "preaching." He said he would leave the decision to Einstein as to the topic of the statement, he hailed the prophetic spirit of Einstein, and ended by calling down God's blessing upon him.

Here is the statement that Einstein sent. It was sent in English on 20 November 1950:

> The most important human endeavor is the striving for morality in our actions. Our inner balance and even our very existence depend on it. Only morality in our actions can give beauty and dignity to life.
>
> To make this a living force and bring it to clear consciousness is perhaps the foremost task of education.
>
> The foundation of morality should not be made dependent on myth nor tied to any authority lest doubt about the myth or about the legitimacy of the authority imperil the foundation of sound judgment and action.

On 27 January 1947 Einstein received a telegram from the National Conference of Christians and Jews somewhat peremptorily saying that it needed a statement from Einstein of 25 to 50 words, to be wired collect, the statement to be in support of "American Brotherhood." It is a topic that invites and almost demands platitudes, but Einstein avoided the pitfall. He sent the following statement in English:

> If the believers of the present-day religions would earnestly try to think and act in the spirit of the founders of these religions then no hostility on the basis of religion would exist among the followers of the different faiths. Even the conflicts in the realm of religion would be exposed as insignificant.

On 14 October of the same year, Einstein received a long telegram saying that on 19 October many diplomats and other distinguished figures would be speaking at a great and impressive dedication of the site on Riverside Drive in New York City on which there would later be erected a memorial to the heroes of the battle of the Warsaw Ghetto and to the six million martyred Jews of Europe. Einstein was invited to attend as an honored guest. If he could not do so, the telegram went on to say, then perhaps he could honor the occasion by sending a message by 16 October.

Einstein needed no urging. It was a cause close to his heart. He sent the following message in English, dating it 19 October 1947:

> Today's solemn meeting has deep significance. Few years separate us from the most horrible mass crime that modern history has to relate; a crime committed not by a fanaticized mob, but in cold calculation by the government of a powerful nation. The fate of the surviving victims of German persecution bears witness to the degree to which the moral conscience of mankind has weakened.

Today's meeting shows that not all men are prepared to accept the Horror in silence. This meeting is inspired by the will to secure the dignity and the natural rights of individual man. It stands for the recognition of the fact that a tolerable existence for man—and even his bare existence—is tied to our adherence to the eternal moral demands.

For this stand I wish to express my appreciation and thanks as a human being and as a Jew.

o

ON 3 August 1946 the Chief Engineer on an American cargo boat wrote a charming letter to Einstein in America telling of an incident on board. The Boatswain and the Carpenter had found a half-starved kitten ashore in Germany, taken it on board, adopted it, and fed it rich food in abundance so that it filled out and flourished and became much attached to its foster parents. But it scratched a seaman who tried to play with it, and he cried out that the cat was crazy. The Boatswain defended the reputation of the kitten saying it was crazy like Einstein, who had had the good sense to leave Germany for the United States. As a result, the kitten was formally given the name "Professor Albert Einstein" by sailors who could not distinguish "relativity" from "kinship."

On 10 August 1946, Einstein replied in English as follows:

> Thank you very much for your kind and interesting information. I am sending my heartiest greetings to my namesake, and also from our own tomcat who was very interested in the story and even a little jealous. The reason is that his own name "Tiger" does not express, as in your case, the close kinship to the Einstein family.
>
> With kind greetings to you, to my namesake's foster parents, and to my namesake himself, . . .

Here are two letters that Einstein sent from Princeton to Gertrud Warschauer in England. She was the widow of a Berlin rabbi, and the letters thank her for Christmas presents that she had sent him on two successive years. A word as to the Englishman, Michael Faraday, mentioned in the second letter: He was a self-taught genius and a lovable man, and one of the greatest experimental physicists of all time. His discoveries and revolutionary ideas in the field of electromagnetism were crucial to the development of relativity. The first letter is dated 2 January 1952:

Dear Gertrud,

The cute ruler that you sent me lies before me. Up to now it was left to intuition to decide whether something that I produced was straight or crooked, parallel, or oblique. I see, however, that, if possible, you would rather avoid being in the hands of the gods (that's how I interpret the ruler).

The second letter was dated 27 December 1952:

You have given me great joy with the little book about Faraday. This man loved mysterious Nature as a lover loves his distant beloved. In his day there did not yet exist the dull specialization that stares with self-conceit through hornrimmed glasses and destroys poetry. . . .

Here is a translation of a quatrain that was found among Einstein's papers. There is no indication of the date or the occasion. It does not seem to have been published before:

That little word "WE" I mistrust, and here's why:
No man of another can say "He is I."
Behind all agreement lies something amiss.
All seeming accord cloaks a lurking abyss.

In reading the following, it is well to bear two things in mind. One is that when Titian was painting a portrait of the Emperor Charles V he dropped a paint brush, which the Emperor graciously picked up for him, saying that Titian deserved to be served by an Emperor. The other is that St. Florian was often depicted as carrying a vessel from which he poured flames, and that his protection was often invoked against fire. However, the phrase used by Einstein in his postscript is a German catch phrase that is applied to a wide assortment of calamities.

A famous German author had been sketching portraits of famous people for ultimate publication in a book. He had just received a cable from an American magazine asking him to make a portrait of Einstein, which, after it appeared in the magazine, he planned to include in his book. Accordingly, on 12 November 1931 he wrote a persuasive letter to Einstein in Berlin asking him if he would consent to pose for his portrait. He said that when he approached politicians they always consented because they needed the publicity, but he well realized that Einstein would be reluctant. He added that even the Emperor Charles V had posed a few times for Titian, and he promised that, in view of the different ratios of greatness on both sides, he would not require Einstein to pick up his paint brush.

On 17 November 1931 Einstein replied as follows:

Do you really believe that the Emperor Charles V would have been so enthusiastic if Titian had painted a picture postcard of him that every Tom, Dick, and Harry could obtain for 10 Pfennig? I believe that he would have picked up the brush for Titian no less cheerfully, but would surely have asked Titian to spare him such publicity—at least during his lifetime.

So please do not be angry with me if I, too, feel that way. Besides, I have to leave for California in a few days, and my hands are full. . . .

P.S. O St. Florian, spare my house. Burn some other fellow's down!

A scientific conference was planned for 1955 in Bern to celebrate the fiftieth anniversary of the special theory of relativity, which Einstein had conceived while working in the patent office there. His friend Max von Laue wrote inviting him to attend as the guest of honor. But Einstein was now in his mid-seventies, with death about to claim him. He answered von Laue in February 1955 with these words:

Old age and ill health make it impossible for me to take part in such occasions, and also I must confess that this divine dispensation is somewhat liberating for me. For everything that has anything to do with the cult of personality has always been painful to me.

The following is taken from a letter to an artist friend written on 27 December 1949:

> It is really a puzzle what drives one to take one's work so devilishly seriously. For whom? For oneself?—one soon leaves, after all. For one's contemporaries? For posterity? *No*, it remains a puzzle.

O

EINSTEIN'S fiftieth birthday, 14 March 1929, was a major event, with gifts and messages of congratulation pouring in from all over the world and reporters of all sorts seeking interviews. Fearing something of the sort, Einstein had fled his Berlin apartment and gone into hiding. When it was over, Einstein was faced with the problem of thanking the many friends who had sent him birthday greetings. He solved it by composing a rhyme, of which the following is a translation, having a printer prepare copies of his handwritten manuscript, and sending these holograph copies to his friends—often with a brief personal greeting added:

Everyone's greeting me today
In the nicest possible way.
Heartfelt words from far and near
Have come from people I hold dear;
And presents, too, to satisfy
Even a gourmet such as I.
They're doing all one possibly can
To satisfy an aged man.
In tones like sweetest melody
They beautify the day for me.
Even cadgers and their pals
Dedicate their madrigals.
So I feel lifted up on high
Like the stately eagle in the sky.
Now the long day nears its end
And greetings to you all I send.

All that you did on my behalf
Has caused the lovely sun to laugh.

A. Einstein
peccavit* 14 March 1929

Among the letters that Einstein received on the occasion of his fiftieth birthday was one from the Nobel laureate Fritz Haber. Here is an excerpt from it:

> "Centuries from now the man in the street will know of our time as the period of the [First] World War, but the educated man will associate the first quarter of our century with your name, just as today some people think of the end of the seventeenth century as the time of the wars of Louis XIV and others as the time of Newton."

A decade later, for Einstein's sixtieth birthday, the Nobel laureate Max von Laue, to whom Einstein felt close, sent the following (Einstein was now in Princeton):

> ". . . Now indeed you are safe, and beyond the reach of that hatred. As I know you, you have come to terms with it within and you stand *above* your fate. But, more than ever, your work is and remains beyond the reach of passion of any kind,

* Artists used to write on their canvases the word "fecit," which is the Latin for "he made [it]," and would then write their name and the date. Here Einstein uses the word "peccavit," which is the Latin for "he sinned [it]."

and will endure as long as there exists a civilized community on this earth."

On 1 May 1936 a prominent American publisher wrote to ask Einstein a favor. The publisher had just broken ground for a new library wing for his country home and wanted to place in the cornerstone an airtight metal box containing items that would be of archeological interest to posterity. There would be, for example, an issue of the *New York Times* specially printed on long-lasting rag paper. He asked Einstein to contribute a message and enclosed, for Einstein to write it on, a sheet of coupon bond paper made out of rag stock that, he assured Einstein, was expected to last a thousand years.

On 4 May 1936 Einstein sent the following message, presumably typed on the special, long-lasting paper:

> Dear Posterity,
>
> If you have not become more just, more peaceful, and generally more rational than we are (or were)—why then, the Devil take you.
>
> Having, with all respect, given utterance to this pious wish,
>
> I am (or was),
> Your,
> Albert Einstein

A correspondent asked Einstein two questions. The first was whether he owed anything to so-called speculative philosophy. The second, which was rather rambling, was whether Einstein thought also that, with current physical researches into space, time, causality, the boundaries of the universe, Beginning, and End, this "science," i.e. speculative philosophy, had become unemployed, or, quoting the scientist R. C. Tolman, whether Einstein agreed that "philosophy is the systematic misuse of a terminology specially invented for that purpose."

On 28 September 1932 Einstein responded from Berlin as follows:

> Philosophy is like a mother who gave birth to and endowed all the other sciences. Therefore one should not scorn her in her nakedness and poverty, but should hope, rather, that part of her Don Quixote ideal will live on in her children so that they do not sink into philistinism.

In 1957 a correspondent, learning that the Estate of Albert Einstein would welcome relevant material, wrote telling of a proposed television program seven years earlier that was entitled "How I Would Spend the Last Two Minutes." Presumably each person interviewed would somehow know that the two minutes in question were indeed going to be those immediately preceding death. Distinguished people such as Eleanor Roosevelt and Albert Schweitzer were to be approached, and the correspondent sent an invitation to Einstein. The topic sounds fascinating–but only at first glance. Einstein saw more deeply. Here is Einstein's response, written in English and dated 26 August 1950:

I feel unable to participate in your projected Television Broadcast "The last two minutes." It seems to me not so relevant how people are to spend the last two minutes before their final deliverance.

In telling of this, the correspondent added the remark: "Needless to say this considerably changed my life."

Einstein was notoriously unconcerned about his attire, which was often quite sloppy. In 1955, early in March, the children in a fifth grade class in an elementary school in the state of New York became aware not only of Einstein's existence but also of the fact that his seventy-sixth birthday would fall a few days hence. With the help of their teacher, they sent Einstein a letter on 10 March wishing him many happy returns, and with the letter they sent a present consisting of a tie clasp and a set of cuff links. It was to be Einstein's last birthday.

On 26 March 1955 Einstein wrote back in English as follows:

Dear Children,

I thank you all for the birthday gift you kindly sent me and for your letter of congratulation. Your gift will be an appropriate suggestion to be a little more elegant in the future than hitherto. Because neckties and cuffs exist for me only as remote memories.

From a boarding school in Cape Town, South Africa, on 10 July 1946, a British student wrote a long and naively charming letter to Einstein in Princeton asking for his autograph. Here is an excerpt: "I probably would have written ages ago, only I was not aware you were still alive. I am not interested in history, and I thought that you had lived in the eighteenth century or somewhere near that time. I must have been mixing you up with Sir Isaac Newton or someone." The student, mentioning a friend, went on to tell that they were much interested in astronomy and would creep past the prefect's room at night so as to observe the stars and planets despite being caught and punished a few times. The student confessed an inability to understand curved space. And ended by saying with sturdy patriotism, "I am sorry that you have become an American citizen. I would much prefer you in England."

On 25 August 1946 Einstein replied in English as follows:

> Dear Master . . . ,
>
> Thank you for your letter of July 10th. I have to apologize to you that I am still among the living. There will be a remedy for this, however.
>
> Be not worried about "curved space." You will understand at a later time that for it this status is the easiest it could possibly have. Used in the right sense the word "curved" has not exactly the same meaning as in everyday language.
>
> I hope that you and your friend's future astronomical investigations will not be discovered any more by the eyes and ears of your school-govern-

ment. This is the attitude taken by most good citizens towards their government and I think rightly so.

Yours sincerely,

The student was thrilled to receive this autographed letter, even though Einstein had mistakenly thought, from her unusual first name, that she was a boy. In her reply, dated 19 September 1946, she wrote: "I forgot to tell you . . . that I was a girl. I mean I am a girl. I have always regretted this a great deal, but by now I have become more or less resigned to the fact." And later in the letter she said: "I say, I did not mean to sound disappointed about my discovery that you were still alive."

Einstein replied:

I do not mind that you are a girl, but the main thing is that you yourself do not mind. There is no reason for it.

o

THE following item was written by Einstein in Princeton, probably in 1935. On the manuscript appear the words "not published." After Einstein's death it was published by Otto Nathan and Heinz Norden in their book *Einstein on Peace*. It is an unusually vehement statement and perhaps that is why Einstein did not publish it. Writing it, however, must have brought him a feeling of relief:

To the everlasting shame of Germany, the spectacle unfolding in the heart of Europe is tragic and grotesque; and it reflects no credit on the community of nations which calls itself civilized!

For centuries the German people have been subject to indoctrination by an unending succession of schoolmasters and drill sergeants. The Germans have been trained in hard work and made to learn many things, but they have also been drilled in slavish submission, military routine and brutality. The postwar democratic Constitution of the Weimar Republic fitted the German people about as well as the giant's clothes fitted Tom Thumb. Then came inflation and depression, with everyone living under fear and tension.

Hitler appeared, a man with limited intellectual abilities and unfit for any useful work, bursting with envy and bitterness against all whom circumstance and nature had favored over him. Sprung from the lower middle class, he had just enough class conceit to hate even the working class which was struggling for greater equality in living standards. But it was the culture and education which had been denied him forever that he hated most of all. In his desperate ambition for power he discovered that his speeches, confused and pervaded with hate as they were, received wild acclaim from those whose situation and orientation resembled his own. He picked up this human flotsam on the streets and in the taverns and organized them around himself. This is the way he launched his political career.

But what really qualified him for leadership was his bitter hatred of everything foreign and, in particular, his loathing of a defenseless minority, the German Jews. Their intellectual sensitivity left him uneasy and he considered it, with some justification, as un-German.

Incessant tirades against these two "enemies" won him the support of the masses to whom he promised glorious triumphs and a golden age. He shrewdly exploited for his own purposes the centuries-old German taste for drill, command, blind obedience and cruelty. Thus he became the *Fuehrer*.

Money flowed plentifully into his coffers, not least from the propertied classes who saw in him a tool for preventing the social and economic liberation of the people which had its beginning under the Weimar Republic. He played up to the people with the kind of romantic, pseudo-patriotic phrasemongering to which they had become accustomed in the period before the World War, and with the fraud about the alleged superiority of the "Aryan" or "Nordic" race, a myth invented by the anti-Semites to further their sinister purposes. His dis-

> jointed personality makes it impossible to know to what degree he might actually have believed in the nonsense which he kept on dispensing. Those, however, who rallied around him or who came to the surface through the Nazi wave were for the most part hardened cynics fully aware of the falsehood of their unscrupulous methods.

Leo Baeck was the leading Rabbi of the Jewish community in Berlin, and a scholar of world renown. When the Nazis came to power he received many attractive offers of positions outside Germany that would have provided physical escape from the Nazi anti-Semitic terror. He refused them all, choosing to share the dangers with his fellow Jews in Germany. After being arrested several times, he was sent to the concentration camp in Terezin, where he remained until the collapse of the German armies, when he was freed by Russian soldiers.

In May of 1953, in a moving tribute to Leo Baeck on the occasion of his eightieth birthday, Einstein wrote from Princeton:

> What this man meant to his brethren trapped in Germany and facing certain destruction cannot fully be grasped by those whose outer circumstances permit them to live on in apparent security. He felt it an obvious duty to stay and endure in the land of merciless persecution in order to provide spiritual sustenance to his brethren till the end. Heedless of danger, he negotiated with the representatives of a government consisting of vicious

murderers and, in every situation, maintained his own and his people's dignity.

When asked to contribute to a Festschrift in honor of Rabbi Baeck, Einstein replied on 23 February 1953 as follows:

> Wishing to assist your fine undertaking and yet being incapable of producing a contribution in the field of our revered and beloved friend, I hit upon this bizarre idea: to put together in the form of pills something out of my own experience that could give our friend a little pleasure—where, though, only the first pill would be allowed to claim a connection with him.

The "pills," for the most part, turned out to be biting aphorisms, of which the following is a sample:

> In order to be an immaculate member of a flock of sheep, one must above all be a sheep oneself.

The first of the "pills" was addressed to Baeck. It was not an aphorism but an affirmation:

> Hail to the man who went through life always helping, who knew no fear, and to whom all aggressiveness and resentment were alien. From such timber are carved the models on which we pattern ourselves, and in whom mankind finds solace in the midst of suffering of its own making.

On 17 March 1954 Rabbi Baeck sent the following letter to Einstein for Einstein's seventy-fifth birthday:

> "In days when the question of the existence of morality seemed to find only the answer "no," or when the very concept of humanity remained in doubt, I was privileged to think of you, and feelings of peace and affirmation came over me. How many a day have you stood before me in my mind and spoken to me."

Einstein died in Princeton on 18 April 1955. On 26 April 1955 Cornelius Lanczos sent these words to Einstein's daughter Margot:

> ". . . One feels that such a man lives forever, in the sense that a man like Beethoven can never die. But there is something forever lost: his sheer joy of living, which was so much a part of his being. It is hard to realize that this man, so unbelievably modest and unassuming, abides with us here no longer. He was aware of the unique role that Fate had bestowed on him, and aware, too, of his greatness. But precisely because this greatness was so towering, it made him modest and humble–not as a pose but as an inner necessity. . . ."

Early in 1933, Einstein received a letter from a professional musician who presumably lived in Munich. The musician was evidently troubled and despondent, and out of a job, yet, at the same time, he must have been something of a kindred spirit. His letter is lost, all that survives being Einstein's reply. Since that reply was dated 5 April 1933, it was presumably sent from Le Coq. Here it is in part. Its haunting despair is timeless, relieved only by the fact that Einstein himself never gave up the fight against darkness. Note the careful anonymity of the first sentence—the recipient would be safer that way:

> I am the one to whom you wrote in care of the Belgian Academy. . . . Read no newspapers, try to find a few friends who think as you do, read the wonderful writers of earlier times, Kant, Goethe, Lessing, and the classics of other lands, and enjoy the natural beauties of Munich's surroundings. Make believe all the time that you are living, so to speak, on Mars among alien creatures and blot out any deeper interest in the actions of those creatures. Make friends with a few animals. Then you will become a cheerful man once more and nothing will be able to trouble you.
>
> Bear in mind that those who are finer and nobler are always alone—and necessarily so—and that because of this they can enjoy the purity of their own atmosphere.
>
> I shake your hand in heartfelt comradeship,
>
> E.

Einstein was the greatest scientist in the world. But the world was such that he signed the letter with a solitary E and not with

Albert Einstein

GERMAN ORIGINALS

HERE are the German texts of writings that were sent out in German; and also, whenever available, texts of preliminary German drafts of items that were sent out in English. The page references show where the corresponding English versions begin.

PAGE 5

Eines Tages erhielt ich im Berner Patentamt ein grosses Couvert, aus dem ein nobles Papier herauskam, auf dem in pittureskem Druck (ich glaube sogar auf Lateinisch) etwas stand, das mich unpersönlich und wenig interessant anmutete und sofort in den amtlichen Papierkorb flog. Später erfuhr ich, dass dies eine Einladung zur Calvinfeier war nebst Ankündigung, dass ich an der Genfer Universität den Ehrendoktor bekommen sollte. Die Leute dort interpretierten offenbar mein Schweigen richtig und wandten sich an meinen Freund und Schüler, den Genfer Chavan, der in Bern lebte. Dieser überredete mich, nach Genf zu kommen weil dies praktisch unvermeidlich wäre, ohne weitere Erklärung.

So fuhr ich am angesagten Tag ab und traf Abends einige Züricher Professoren im Restaurant des Gasthofes wo wir wohnten. . . . Jeder von ihnen erzählte nun, in welcher Eigenschaft er da war. Als ich schwieg, erging die Frage an mich, und ich musste gestehen, dass ich keine blasse Ahnung habe. Die andern wussten aber Bescheid und weihten mich ein. Am nächsten Tag sollte ich im Festzug marschieren und hatte nur Strohhut und Strassenanzug bei mir. Mein Vorschlag, mich davon zu drücken, wurde mit Entschiedenheit abgelehnt, und diese Feier verlief entsprechend drollig, was meine Mitwirkung anlangte.

Das Fest endete mit dem opulentesten Festessen, dem ich in meinem ganzen Leben beigewohnt habe. Da sagte ich zu einem Genfer Patrizier, der neben mir sass: "Wissen Sie, was Calvin gemacht hätte, wenn er noch da wäre?" Als er verneinte und mich um die Meinung fragte, sagte ich: "Er würde einen grossen Scheiterhaufen errichtet und uns alle wegen sündhafter Schlemmerei verbrannt haben". Der Mann sprach kein Wort mehr und damit endet meine Erinnerung an die denkwürdige Feier.

PAGE 7

Sie können sich kaum vorstellen, wie erfreut ich darüber war und bin, dass die Berner Naturforschende Gesellschaft meiner so freundlich gedacht hat. Es war sozusagen eine Botschaft aus den Tagen der längst entschwundenen Jugend. Die gemütlichen und anregenden Abende steigen wieder auf in meinem Gedächtnis und besonders die oft wunderbaren Bemerkungen, die Professor Sahli [Salis?], der innere Mediziner, zu den Vorträgen zu machen pflegte. Ich habe das Dokument gleich einrahmen lassen und als Einziges von allen Anerkennungen entsprechender Art in meinem Studierzimmer aufgehängt als Erinnerungszeichen an meine Berner Zeit und die dortigen Freunde.

Ich bitte Sie, der Gesellschaft meinen herzlichen Dank zu übermitteln und ihr zu sagen, wie hoch ich die mir erwiesene Freundlichkeit schätze.

PAGE 8

Ich werde nämlich mit der Berühmtheit immer dümmer, was ja eine ganz gewöhnliche Erscheinung ist. Das Missverhältnis zwischen dem, was man ist, und dem, was die andern von einem glauben oder wenigstens sagen, ist gar zu gross. Man muss es aber mit Humor tragen . . .

PAGE 9

Am 14. März 1879 wurde ich in Ulm geboren und gelangte im Alter von einem Jahre nach München, wo ich bis zum Winter 1894-1895 verblieb. Dort besuchte ich die Elementarschule und das Luitpoldgymnasium bis zur 7. Klasse (exklusive). Dann lebte ich bis zum Herbste vorigen Jahres in Mailand, wo ich privatim weiterstudierte. Seit letzten Herbst besuche ich die Kantonsschule in Aarau und erlaube mir nun, mich zur Maturitätsprüfung anzumelden. Ich gedenke dann an der 6. Abteilung des Eidgenössischen Polytechnikums Mathematik und Physik zu studieren.

PAGE 10

I. Ich bin in Ulm als Sohn jüdischer Eltern am 14. März 1879 geboren. Mein Vater war Kaufmann, zog bald nach meiner Geburt nach München, 1893 nach Italien, wo er bis zu seinem Tode (1902) blieb. Ich habe keinen Bruder, aber eine Schwester, die in Italien lebt. . . .

IV. Von 1900 bis 1902 war ich in der Schweiz als Privatlehrer, eine Zeitlang auch als Hauslehrer tätig und erwarb das Schweizerische Bürgerrecht. 1902-1909 war ich als Experte (Vorprüfer) am Eidgenöss. Amt für geistiges Eigentum angestellt, 1909-11 als ausserordentlicher Professor an der Züricher Universität. 1911-12 war ich als ordentlicher Professor der theoretischen Physik an der Universität Prag, 1912-14 an dem Eidgenössischen Polytechnikum ebenfalls als Professor der theoretischen Physik. Seit 1914 bin ich als bezahltes Mitglied an der Preussischen Akademie der Wissenschaften in Berlin und kann mich ausschliesslich der wissenschaftlichen Forschungsarbeit widmen.

PAGE 11

V. Meine Veröffentlichungen bestehen fast ausschliesslich in kurzen physikalischen Arbeiten, welche meist in

den Annalen der Physik und in den Sitzungsberichten der Preuss. Akademie erschienen sind. Die wichtigsten betreffen folgende Themen:
Brown'sche Bewegung (1905)
Theorie der Planck'schen Formel und der Lichtquanten (1905, 1917)
Spezielle Relativitätstheorie und Trägheit der Energie (1906)
Allgemeine Relativitätstheorie 1916 und später
Ferner sind Arbeiten über die thermischen Schwankungen zu erwähnen, sowie eine 1931 [E. schrieb versehentlich 1917] mit Prof. W. Mayer verfasste Arbeit über die einheitliche Natur von Gravitation und Elektrizität.

VI. Gelegentliche Vortragsreisen nach Frankreich, Italien, Japan, Argentinien, England, die Vereinigten Staaten, die–abgesehen von den Reisen nach Pasadena–nicht eigentlich Forschungszwecken dienten.

PAGE 12

VII. Mein eigentliches Forschungsziel war stets die Vereinfachung und Vereinheitlichung des physikalischen theoretischen Systems. Dies Ziel erreichte ich befriedigend für die makroskopischen Phänomene, nicht aber für die Phänomene der Quanten und die atomistische Struktur. Ich glaube, dass auch die moderne Quantenlehre von einer befriedigenden Lösung des letzteren Problemkomplexes trotz erheblicher Erfolge noch weit entfernt ist.

VIII. Ich wurde Mitglied vieler wissenschaftlicher Gesellschaften, und mehrere Medaillen wurden mir verliehen, auch eine Art Gastprofessur an der Universität Leiden. In einer ähnlichen Verbindung stehe ich zur Universität Oxford (Christ Church College).

PAGE 14

"Er pflegte in seiner Jugend oft zu sagen: 'Ich will einmal in meinem Esszimmer nur einen tannenen Tisch, eine Bank und ein paar Stühle.' "

PAGE 14

Am meisten drückt mich natürlich das Unglück meiner armen Eltern. Ferner schmerzt es mich tief, dass ich als erwachsener Mensch untätig zusehen muss, ohne auch nur das Geringste machen zu können. Ich bin ja nichts als eine Last für meine Angehörigen. . . . Es wäre wirklich besser, wenn ich gar nicht lebte. Der einzige Gedanke, dass ich immer alles getan habe, was mir meine kleinen Kräfte erlaubten und dass ich mir jahr-in jahraus auch nicht einmal ein Vergnügen, eine Zerstreuung erlaube, ausser die welche mir das Studium bietet, hält mich noch aufrecht und muss mich manchmal vor Verzweiflung schützen.

PAGE 14

Es gibt ziemlich zu tun doch nicht übermässig, sodass ich schon einmal Zeit finde, in Zürichs schöner Umgebung ein Stündchen zu verbummeln. Zudem freue ich mich bei dem Gedanken, dass nun für meine Eltern die schwersten Sorgen ein Ende haben. Wenn alle Leute so lebten (so wie ich), wahrlich, die Romanschriftstellerei wäre dann niemals auf die Welt gekommen.

PAGE 15

In Berlin alles aufgeben, wo man mir so unbeschreiblich entgegenkommt, das brächte ich nicht fertig. Wie glücklich wäre ich vor 18 Jahren gewesen, wenn ich am Polytechnikum bescheidener Assistent hätte werden können! Aber es gelang mir nicht! Die Welt ist ein Narrenhaus, das Renommee macht alles! Ein anderer liest doch auch gut—aber. . . .

PAGE 16

Mit der Arbeit geht es langsam und harzig, nach einem vielversprechenden Anfang. Wir sind in den prinzipiellen Forschungen in der Physik in einer Situation des Tastens, wo keiner Vertrauen in das hat, was der andere mit grossen Hoffnungen versucht. Man ist in einer beständigen Spannung, bis man endgültig absegelt. Es bleibt mir aber der Trost, dass das Hauptsächliche, was ich gemacht habe, zu dem selbverständlichen Bestande unserer Wissenschaft geworden ist.

Die grossen politischen Dinge unserer Zeit sind so entmutigend, dass man sich in der eigenen Generation ganz vereinsamt fühlt. Es ist, wie wenn die Menschen die Leidenschaft für Recht und Würde verloren hätten und nicht mehr schätzten, was bessere Generationen mit unsäglichem Opfermut erworben haben. . . . Das Fundament aller menschlichen Werte ist eben das Moralische. Das in primitiver Zeit klar gesehen zu haben, ist die einzigartige Grösse unseres Moses. Schau Dir die Heutigen dagegen an!

PAGE 16

Nichts als Briefschulden habe ich und Menschen, die mit Recht unzufrieden mit mir sind. Aber kann es anders sein bei einem Besessenen? Wie in der Jugend sitze ich unaufhörlich da und denke und rechne, hoffend tiefe Geheimnisse zu lüften. Die sogenannte grosse Welt, das heisst das menschliche Getriebe hat weniger Anziehendes als je, sodass man jeden Tag in der Einkapselung bestärkt wird.

PAGE 17

Ich sollte ursprünglich auch Techniker werden. Aber der Gedanke, die Erfindungskraft auf Dinge verwenden zu sollen, welche das werktägliche Leben noch raffinierter machen, mit dem Ziel öder Kapitalschinderei, war mir unerträglich. Das Denken um seiner selbst willen wie die Musik! . . . Wenn ich kein Problem zum Nachdenken habe, dann leite ich mit Vorliebe mathematische und physikalische Sätze wieder ab, die mir längst bekannt sind. Hier ist also gar kein *Ziel* da, sondern nur eine Gelegenheit, um sich der angenehmen Tätigkeit des Denkens hinzugeben.

PAGE 18

Was das Suchen nach Wahrheit anbelangt, so weiss ich aus eigenem mühevollen Suchen und vielem Verzichten, wie schwer es ist, in der Erkenntnis des wirklich Wesentlichen einen zuverlässigen, wenn auch kleinen Schritt zu finden.

PAGE 18

Der theoretisch arbeitende Naturforscher ist nicht zu beneiden, denn die Natur, oder genauer gesagt: das Experiment, ist eine unerbittliche und wenig freundliche Richterin seiner Arbeit. Sie sagt zu einer Theorie nie "ja" sondern im günstigsten Falle "vielleicht", in den meisten Fällen aber einfach "nein". Stimmt ein Experiment zur Theorie, bedeutet es für letztere "vielleicht", stimmt es nicht, so bedeutet es "nein". Wohl jede Theorie wird einmal ihr "nein" erleben, die meisten Theorien schon bald nach ihrer Entstehung.

PAGE 19

Äussere Ereignisse, die an sich fähig wären, die Richtung des Denkens und Handelns einer Person zu bestimmen, kommen wohl in jedem Leben vor. Bei den meisten aber bleiben solche Vorkommnisse ohne Wirkung. In meinem Leben hat sicher der ungeheure Eindruck eine Rolle gespielt, den auf mich ein kleiner Kompass machte, den mir mein Vater zeigte, als ich ein kleiner Junge war. Von der Riemann'schen Arbeit erfuhr ich erst zu einer Zeit, in der die Grundprinzipien der allgemeinen Relativitäts-Theorie schon längst klar konzipiert waren.

PAGE 20

Es hat mich etwas befremdet, dass Sie bezüglich des Zusammenhanges zwischen träger Masse und Energie meine Priorität nicht anerkennen.

PAGE 20

Wenn es mir schon vor Empfang Ihres Briefes leid tat, dass ich mir durch eine kleinliche Regung jene Äusserung über Priorität in der bewussten Sache diktieren liess, zeigte mir Ihr ausführlicher Brief erst recht, dass meine Empfindlichkeit übel angebracht war. Die Leute, denen es vergönnt ist, zum Fortschritt der Wissenschaft etwas beizutragen, sollten sich die Freude über die Früchte gemeinsamer Arbeit nicht durch solche Dinge trüben lassen.

PAGE 21

Ich bin nicht dafür, dass es abgedruckt wird, weil der Vortrag nicht originell genug ist. Man muss vor allem kritisch gegen sich selbst sein. Den Anspruch gelesen zu werden, kann man nur dann aufrecht erhalten, dass man alles Unbedeutende verschweigt, wenn es irgend geht.

PAGE 21

Ihr heiliger Eifer in Bezug auf die Kritik im Literary Supplement der [London] Times hat bei mir gutmütige Heiterkeit hervorgerufen. Da schreibt einer für wenig Geld auf Grund flüchtigen Durchblätterns etwas, was einigermassen plausibel klingt und was von niemand genau gelesen wird. Wie können Sie sich darüber ernsthafte Gedanken machen. Über mich sind schon massenweise so unverschämte Lügen und freie Erfindungen von Reportern erschienen, dass ich längst unterm Boden wäre, wenn ich mich darum kümmern wollte. Man muss sich damit trösten, dass die Zeit ein Sieb hat, durch welches die meisten Wichtigkeiten ins Meer der Vergessenheit ablaufen und was bei dieser Auslese übrig bleibt ist oft immer noch fad und schlecht. . . .

PAGE 22

Jeder Piepser wird bei mir zum Trompetensolo . . .

PAGE 22

Früher dachte ich nicht daran, dass jedes spontan geäusserte Wort aufgegriffen und fixiert werden könnte. Sonst hätte ich mich mehr ins Schneckenhaus verkrochen.

PAGE 22

In England sind sogar die Reporter zurückhaltend! Ehre, wem Ehre gebührt. Einmaliges Nein genügt. Die Welt kann da noch viel lernen–nur ich will es nicht und kleide mich stets nachlässig, auch bei dem heiligen Sakrament des Dinners.

PAGE 23

Stilles Dasein in der Klause mit argem Frieren. Abends das feierliche Abendmahl der heiligen Brüderschaft im Frack.

PAGE 23

Solchen Sturm wie in dieser Nacht habe ich . . . noch nicht mit gemacht. . . . Der Anblick des Meeres ist unbeschreiblich grossartig, besonders wenn Sonne darauf fällt.

Man ist wie aufgelöst in die Natur. Man fühlt die Belanglosigkeit des Einzelgeschöpfes noch mehr als sonst und ist froh dabei.

PAGE 24

Am Objektiven gemessen ist es unsäglich wenig, was der Mensch durch heisses Bemühen der Wahrheit abringt. Aber das Streben befreit uns aus den Fesseln des Ichs und macht uns zu Genossen der Besten.

PAGE 24

Wie armselig steht der theoretische Physiker vor der Natur und vor–seinen Studenten!

PAGE 28

Ich muss offen sagen, dass ich es nicht billige, wenn Eltern auf die Entschliessungen ihrer Kinder Einfluss nehmen, die für die Gestaltung ihres Lebens entscheidend sind. Solche Probleme muss jeder für sich selbst lösen.

Wenn Sie aber eine Entscheidung treffen wollen, mit der Ihre Eltern nicht einverstanden sind, so müssen Sie sich fragen: bin ich innerlich unabhängig genug, um entgegen dem Willen der Eltern handeln zu können, ohne dabei mein inneres Gleichgewicht zu verlieren? Wenn Sie dessen nicht sicher sind, so ist der Schritt auch im Interesse des Mädchens nicht zu empfehlen. Davon allein sollten Sie Ihre Entscheidung abhängig machen.

PAGE 29

Zur Aufstellung einer Theorie genügt niemals das blosse Zusammenbringen registrierter Erscheinungen–es muss stets eine freie Erfindung des menschlichen Geistes hinzukommen, die dem Wesen der Dinge näher auf den Leib rückt. Und: der Physiker darf sich nicht begnügen mit der blossen phänomenologischen Betrachtung, die nach der Erscheinung frägt, sondern muss vordringen zur spekulativen Methode, welche die Existenzform sucht.

PAGE 30

Jugend, weisst du, dass du nicht die erste Jugend bist, die nach einem Leben voll Schönheit und Freiheit lechzte?

Jugend, weisst du, dass all deine Vorfahren so waren wie du und der Sorge und dem Hass verfielen?

Weisst du auch, dass deine heissen Wünsche nur dann in Erfüllung gehen können, wenn es dir gelingt, Liebe und Verständnis für Mensch, Tier, Pflanze und Sterne zu erringen, wenn jede Freude deine Freude und jeder Schmerz dein Schmerz sein wird? Öffne deine Augen, dein Herz und deine Hände und meide das Gift, das deine Ahnen aus der Geschichte gierig gesogen haben. Dann wird die Erde dein Vaterland sein und all dein Schaffen und Wirken wird Segen spenden.

PAGE 31

Liebe Kinder,

Wir sollen nicht fragen: "Was ist ein Tier", sondern: "Was für eine Art von Dingen nennen wir ein Tier?" Nun, wir nennen etwas ein Tier, wenn es gewisse Eigenschaften hat. Es nimmt Nahrung auf, stammt von etwa gleich beschaffenen Eltern, es wächst, es bewegt sich selber, es stirbt, wenn seine Zeit vorüber ist. So nennen wir den Wurm, das Huhn, den Hund, den Affen ein Tier. Wie mit uns Menschen? Lasst uns in solcher Art darüber denken und selber entscheiden, ob es natürlich ist, wenn wir uns selbst als Tiere betrachten.

PAGE 32

Ich habe Deine Frage so einfach zu beantworten gesucht als es mir möglich war. Hier meine Antwort:

Der wissenschaftlichen Forschung liegt der Gedanke zugrunde, dass alles Geschehen durch Naturgesetze bestimmt sei, also auch das Handeln der Menschen. Deshalb wird ein Forscher kaum geneigt sein, zu glauben, dass das Geschehen durch ein Gebet–das heisst, durch einen gegenüber einem übernatürlichen Wesen geäusserten Wunsch–beeinflusst werden könnte.

Allerdings muss zugegeben werden, dass unsere tatsächliche Kenntnis dieser Gesetze nur unvollkommenes Stückwerk ist, sodass letzten Endes die Überzeugung von der Existenz letzter durchgreifender Gesetze ebenfalls auf einer Art Glauben beruht. Immerhin ist dieser Glaube weitgehend durch die bisherigen Erfolge der Wissenschaften gerechtfertigt.

Andererseits erfüllt aber die Wissenschaft Jeden, der sich ernsthaft mit ihr befasst, mit der Überzeugung, dass sich in der Gesetzmässigkeit der Welt ein dem menschlichen ungeheuer überlegener Geist manifestiere, demgegenüber wir mit unseren bescheidenen Kräften demütig zurückstehen müssen. So führt die Beschäftigung mit der Wissenschaft zu einem religiösen Gefühl besonderer Art, welches sich von der Religiosität des naiveren Menschen allerdings wesentlich unterscheidet.

Freundlich grüsst Dich

PAGE 33

Liebe Kinder!

Mit Freude stelle ich mir vor, dass Ihr Kinder zu frohem Feste vereinigt seid beim Glanze der Weihnachtslichter. Denket auch an die Lehre dessen, dessen Geburt Ihr durch dieses Fest feiert. Diese Lehre ist so einfach; und doch hat sie sich in fast 2000 Jahren unter den Menschen nicht durchsetzen können: Lernet froh zu sein durch das Glück und die Freude Eurer Genossen, nicht durch den öden Kampf der Menschen untereinander! Wenn Ihr diesem natürlichen Fühlen in Euch Raum gebt, wird Euch jegliche Bürde dieses Lebens leicht oder doch erträglich werden, und Ihr werdet in Gelassenheit und ohne Furcht Euren Weg finden und überall Freude verbreiten.

PAGE 35

Ich bedauere, Ihrem Wunsche nicht entsprechen zu können, weil ich gerne im Dunkel des Nicht-Analysiertseins verbleiben möchte.

PAGE 35

Verehrter Meister!

Ich danke Ihnen herzlich, dass Sie meiner gedacht haben. Warum betonen Sie bei mir das Glück? Sie, der Sie in die Haut so vieler Menschen, ja der Mensch*heit* geschlüpft sind, hatten doch keine Gelegenheit, in die meine zu schlüpfen!

In aller Hochschätzung und mit herzlichen Wünschen

PAGE 37

Ohne das Bewusstsein, dass ich irgendetwas Originelles oder gar der Veröffentlichung Würdiges zu dem von Ihnen genannten Thema zu sagen hätte, sende ich Ihnen die beiliegende aphoristische Äusserung, um Ihnen meinen guten Willen zu beweisen. Wenn meine Tinte weniger dickflüssig wäre, würde ich dem in Ihrem freundlichen Brief geäusserten Wunsche durch ein üppigeres Opus gerecht geworden sein.

PAGE 37

Das Gemeinsame am künstlerischen und wissenschaftlichen Erleben.

Wo die Welt aufhört, Schauplatz des persönlichen Hoffens, Wünschens und Wollens zu sein, wo wir uns ihr als freie Geschöpfe bewundernd, fragend, schauend gegenüberstellen, da treten wir ins Reich der Kunst und Wissenschaft ein. Wird das Geschaute und Erlebte in der Sprache der Logik nachgebildet, so treiben wir Wissenschaft, wird es durch Formen vermittelt, deren Zusammenhänge dem bewussten Denken unzugänglich, doch intuitiv als sinnvoll erkannt sind, so treiben wir Kunst. Beiden gemeinsam ist die liebende Hingabe an das Überpersönliche, Willensferne.

PAGE 38

Körper und Seele sind nicht zwei verschiedene Dinge, sondern nur zwei verschiedene Arten, dasselbe Ding wahrzunehmen. Entsprechend sind Physik und Psychologie nur zwei verschiedenartige Versuche, unsere Erlebnisse auf dem Weg systematischen Denkens miteinander zu verknüpfen.

PAGE 38

Politik ist ein durch beständig verjüngte Illusionen beseeltes Pendeln zwischen Anarchie und Tyrannei.

PAGE 39

Das Missverständnis hier kommt durch eine schlechte Übersetzung eines deutschen Textes, insbesondere der Gebrauch des Wortes "mystical". Ich habe nie der Natur eine Absicht oder einen Zweck zugeschrieben, überhaupt nichts anthropomorphisch zu Deutendes.

Was ich in der Natur sehe, ist eine grossartige Struktur, die wir nur sehr unvollkommen zu erfassen vermögen und die einen vernünftigen Menschen mit einem Gefühl von "Humility" erfüllen muss. Dies ist ein echt religiöses Gefühl, das nichts mit Mystizismus zu schaffen hat.

PAGE 40

Der mystische Zug unserer Zeit, welcher sich besonders in dem Wuchern der sogenannten Theosophie und dem Spiritismus zeigt, ist für mich nur ein Symptom von Schwäche und Zerfahrenheit.

Da unsere seelischen Erlebnisse in Reproduktionen und Kombinationen sinnlicher Eindrücke bestehen, so scheint mir die Konzeption einer Seele ohne Körper leer und nichtssagend.

PAGE 40

Nach meiner Ansicht ist die Situation die folgende:

Die meisten Bücher über Wissenschaft, die für den Laien Verständlichkeit beanspruchen, gehen mehr darauf aus, den Leser zu beeindrucken ("awe-inspiring") (wie weit haben wir es gebracht!) als ihm die elementaren Ziele und Methoden klar zu machen. Wenn der intelligente Laie ein paar solche Bücher in die Hand bekommen hat, so wird er völlig entmutigt. Sein Ergebnis ist: "Ich bin zu schwachköpfig und muss es aufgeben". Dazu kommt, dass die ganze Darstellung meist sensationell ist, was ebenfalls den verständigen Laien abstösst.

Mit einem Wort: die Schuld liegt nicht bei den Lesern sondern bei den Autoren und Verlegern. Mein Vorschlag: Kein solches Buch sollte gedruckt werden, bevor festgestellt wird, dass es von einem intelligenten, kritischen *Laien* verstanden und geschätzt wird.

PAGE 41

Ich bin nicht schuld, wenn Laien eine so übertriebene Vorstellung von der Bedeutung meiner Bemühungen erhalten. Dies kommt vielmehr von den Popularisatoren und ganz besonders von den Zeitungskorrespondenten her, die alles nach Möglichkeit in einem sensationellen Lichte darstellen.

PAGE 44

Liebes Frl. Ney,

Von Elsa höre ich, dass Du nicht zufrieden bist, weil Du den Onkel Einstein nicht gesehen hast. Ich sage Dir daher, wie ich aussehe: Bleiches Gesicht, lange Haare und eine Art bescheidenes Bäuchlein. Dazu ein eckiger Gang und eine Zigarre im Maul, wenn er eine hat, und einen Federhalter in der Tasche oder in der Hand. Krumme Beine und Warzen hat er aber nicht, ist also ganz hübsch, auch keine Haare an den Händen wie oft hässliche Männer. Also doch schade, dass Du mich nicht gesehen hast.

Sei herzlich gegrüsst von
Deinem Onkel Einstein.

PAGE 45

So wie's nun ist, fühl' ich mich recht verlegen,
Wär' ich ein Pfaff' gäb' ich gern meinen Segen!

So freue ich mich, von Ihnen gehört zu haben, sowie davon, dass sich Ihr Sohn der Physik widmen will. Ich kann aber nicht verschweigen, dass dies eine harte Sache ist, wenn man sich nicht mit oberflächlichen Resultaten zufrieden geben will. Am besten scheint es mir, inneres Streben und Metier zu trennen, soweit es möglich ist. Es ist nicht gut, wenn das tägliche Brot an Gottes aussergewöhnlichen Segen gebunden ist.

PAGE 46

Alles wirklich Wertvolle kommt nicht aus dem Ehrgeiz oder aus dem Pflichtgefühl, sondern aus der Liebe und Devotion gegenüber Menschen oder objectiven Dingen.

PAGE 47

. . . wo ich mit rührender Herzlichkeit aufgenommen wurde. Diese beiden Leutchen sind von einer Reinheit und Güte, die selten zu finden ist. Erst unterhielten wir uns eine Stunde. Dann kam eine englische Musikantin und wir musizierten zu viert, bezw. zu dritt (eine musikalische Hofdame war auch noch da) ein paar Stunden lang und sehr vergnügt. Dann gingen alle weg, und ich war allein bei Königs zum Essen; ohne Bedienung, vegetarisch. Spinat mit Setzei und Kartoffeln, punktum. (Es war nicht vorausberechnet, dass ich blieb). Es gefiel mir dort über die Massen und ich bin sicher, dass dieses Gefühl gegenseitig ist.

PAGE 48

Es war eine grosse Freude für mich, Ihnen von den Mysterien zu erzählen, vor die uns die Physik stellt. Man hat als Mensch gerade noch so viel Verstand mitbekommen, dass man von seiner intellektuellen Ohnmacht dem Seienden gegenüber eine deutliche Vorstellung erlangen kann. Die Welt des Menschengetriebes würde schöner aussehen, wenn diese Demut allen mitgeteilt werden könnte.

PAGE 48

Zwei Tage weilte ich in dieser Ecke der Sorglosen, wo Wind, Hitze und Kälte unbekannt sind. Da zeigte man mir gestern eine verträumte Villa (Bliss), in welcher Sie vor ein paar Jahren einige frohe und stille Tage verbracht haben sollen.

Schon zwei Monate bin ich in diesem Lande der Gegensätze und Überraschungen, wo man abwechselnd bewundern und kopfschütteln muss. Man merkt, dass man an dem alten Europa mit seinen Nöten und Schmerzen hängt und kehrt gerne wieder zurück.

Sie und Ihren Herrn Gemahl grüsst in frohem Gedenken der schönen in Brüssel verbrachten Stunden

PAGE 48

Ein Baum im Klostergarten stand
Der war gepflanzt von Ihrer Hand.
Ein Zweiglein sendet er zum Gruss,
Weil er dort stehen bleiben muss.

PAGE 49

Das Zweiglein brachte mir den Gruss
Vom Baum der stehen bleiben muss,

Auch von dem Freund der ihn gepflückt,
Und mich damit so sehr beglückt.
Ich rufe Dank viel tausend Mal
Übers Meer und Berg und Tal,
Und wünsch', da alle Steine wanken jetzt,
Ein Stein doch bleibe unverletzt!

PAGE 49

In der Hauptstadt stolzer Pracht
Wo das Schicksal wird gemacht
Kämpfet froh ein stolzer Mann
Der die Lösung schaffen kann.

Beim Gespräche gestern Nacht
Herzlich Ihrer ward gedacht
Was berichtet werden muss
Darum send' ich diesen Gruss.

PAGE 51

Verehrte Königin,

Heute ist zum ersten Mal in diesem Jahr die Frühlingssonne erschienen und weckte mich aus dem gleichmässigen Traumzustand, in den die wissenschaftliche Arbeit unsereinen versetzt. Da steigen Gedanken herauf vom früheren farbigeren Leben und auch von schönen Stunden in Brüssel.

Frau Barjansky schrieb mir, wie schwer Sie am Leben leiden und wie sehr Sie durch das unsäglich Schmerzliche, das Sie erlitten haben gelähmt sind.

Und doch sollten wir diejenigen nicht beklagen, die nach glücklichem fruchtbaren Wirken in der Blüte ihrer Kraft von uns gegangen sind und denen es vergönnt war, in vollem Masse ihre Lebensaufgabe zu erfüllen.

Was den älteren Menschen erfrischend beleben kann, ist die Freude am Treiben der Jungen, eine Freude, die in diesen verworrenen Zeiten allerdings durch bange Ahnungen verdüstert wird. Und doch erweckt die Sonne des Frühlings neues Leben wie früher und wir dürfen uns darüber freuen und zur Entfaltung dieses Lebens beitragen, und Mozart ist so schön und zart geblieben, wie er immer war und sein wird. Es gibt doch etwas Ewiges, das der Hand des Schicksals und aller menschlichen Verblendung entrückt ist. Und diese ewigen Dinge stehen dem älteren Menschen näher als dem zwischen Furcht und Hoffnung pendelnden jüngeren Menschen. Uns bleibt es vorbehalten, das Schöne und Wahre am reinsten zu erleben.

Haben Sie je die Maximen von La Rochefoucauld gelesen; sie scheinen sehr herb und düster, bringen aber durch ihre Objektivierung der menschlichen und allzumenschlichen Natur eine seltsame Befreiung. Da hat sich einer frei gemacht, der es nicht leicht gehabt hat, das schwere Gepäck von Leidenschaft loszuwerden, das ihm die Natur auf den Lebensweg mitgab. Am hübschesten liest sich dies mit Menschen zusammen, deren Schifflein durch manchen Sturm gegangen ist, z.B. mit den guten Barjanskys. Ich täte auch gern mit, wenn es das grosse Wasser nicht verbieten würde.

Es ist mir vergönnt, hier in Princeton auf einer Schicksalinsel zu leben, die in mancher Beziehung Ähnlichkeit hat mit dem lieblichen Schlossgarten in Laeken. Auch hierher in dies kleine Universitätsstädtchen dringen kaum die wirren Stimmen des menschlichen Kampfes. Ich schäme mich fast in solcher Ruhe zu leben, während sonst alles kämpft und leidet. Aber schliesslich ist es doch am besten, sich um die ewigen Dinge zu bemühen; denn von ihnen allein strömt jener Geist aus, der der Menschenwelt Ruhe und Freude zurückbringen kann.

Indem ich von Herzen hoffe, dass der Frühling auch Ihnen stille Freude bringe und Sie zu frohem Tun anrege, grüsst Sie mit besten Wünschen.

PAGE 54

Sie fragen mich, was ich dachte, als ich davon erfuhr, dass die Potsdamer Polizei mein Sommerhaus überfiel, um da nach verborgenen Waffen zu suchen.

Es fiel mir das deutsche Sprichwort ein: Jeder nimmt das Mass nach seinen eigenen Schuhen.

PAGE 55

In solchen Zeiten hat man Gelegenheit, seine wirklichen Freunde kennen zu lernen. Ich danke Ihnen herzlich für Ihre Hilfsbereitschaft. In Wahrheit geht es mir aber sehr gut, sodass ich nicht nur mit den Meinen durchkommen sondern noch andere über Wasser halten kann. Aus Deutschland werde ich allerdings wohl kaum etwas retten, weil man ein Verfahren wegen Hochverrates gegen mich angestrengt hat. Der Physiologe [Jaques] Löb sagte mir gesprächsweise einmal, dass die politischen Führer eigentlich alle Pathologen sein müssten, da ein normaler Mensch keine so ungeheure Verantwortung tragen könne, wenn er die Konsequenz seiner Entschlüsse und Handlungen so wenig überschauen könne. Wenn dies auch etwas übertrieben sein mag, so trifft es doch gegenwärtig in Deutschland in vollem Masse zu. Kurios ist nur das vollständige Versagen der sogenannten geistigen Aristokratie.

PAGE 56

Sich verlieben ist gar nicht das Duemmste, was der Mensch tut–die Gravitation kann aber nicht dafür verantwortlich gemacht werden.

PAGE 56

Ich freue mich, dass ich unter Euch jungem und frohem Volk leben werde. Wenn Euch ein alter Student etwas in kurzen Worten sagen kann, so ist es dies: Sehet im Studium nie eine Pflicht sondern die beneidenswerte

Gelegenheit, die befreiende Schönheit auf dem Gebiet des Geistes kennen zu lernen zu Eurer eigenen Freude und zugunsten der Gemeinschaft, der Euer späteres Wirken gehört.

PAGE 57

Wissenschaft ist eine wunderbare Sache, wenn man nicht davon leben muss. Den Lebensunterhalt verdienen durch eine Arbeit, von der man sicher ist, dass man sie zu tun fähig ist. Dann allein kann man sich freuen an wissenschaftlichem Streben, für das man niemand Rechenschaft schuldet.

PAGE 60

Ich bin gottlob abseits und brauche mich nicht mehr am Wettrennen der Geister zu beteiligen. Eine Beteiligung daran ist mir immer als schlimme Sklaverei erschienen, nicht weniger als die Sucht nach Geld oder Macht.

PAGE 60

Der Wert des Judentums liegt ausschliesslich in seinem geistigen und ethischen Gehalt und in demjenigen der einzelnen Juden. Deshalb war das Studium mit Recht von Alters her die geheiligte Bemühung der Fähigen unter uns. Dies will aber nicht sagen, dass wir von geistiger Arbeit unseren Lebensunterhalt bestreiten sollen, wie es unter uns leider allzuviel der Fall ist. Wir müssen in dieser schweren Zeit alles daran setzen, eine Anpassung an das praktisch Nötige zu finden, ohne von der Liebe zum Geistigen und der Pflege der Studien abzulassen.

PAGE 61

Im letzten Grunde ist jeder ein Mensch, gleichgültig ob Amerikaner, Deutscher, Jew or Gentile. Wenn es möglich wäre, mit diesem allein würdigen Standpunkt auszukommen, wäre ich ein glücklicher Mensch. Ich finde es traurig, dass im heutigen praktischen Leben Trennungen nach Staatszugehörigkeit und kultureller Tradition eine so erhebliche Rolle spielen. Da dies nun aber einmal unabänderlich ist, darf man sich der Wirklichkeit gegenüber nicht verschliessen.

Was nun die eine alte Traditionsgemeinschaft bildende Judenheit anbelangt, so lehrt deren Leidensgeschichte, dass–mit den Augen des Historikers gesehn–das Jude-Sein sich in stärkerem Masse ausgewirkt hat als die Zugehörigkeit zu politischen Gemeinschaften. Wenn zum Beispiel die deutschen Juden aus Deutschland vertrieben werden, so hören sie auf, Deutsche zu sein, ändern ihre Sprache und ihre politische Zugehörigkeit, aber sie bleiben Juden. Warum dies so ist, ist gewiss eine schwierige Frage; ich sehe den Grund in der Hauptsache nicht in Merkmalen der Rasse, sondern in fest eingewurzelten Traditionen, die sich keineswegs auf das Religiöse beschränken. An dieser Tatsache wird dadurch nichts geändert, dass Juden als Bürger bestimmter Staaten in deren Kriegen zum Opfer fallen.

PAGE 62

Samstag gehts auf die Rutsch nach Amerika, und zwar nicht um an Universitäten zu sprechen (was wohl nebenbei auch geschehen wird), sondern wegen der Gründung der jüdischen Universität in Jerusalem. Ich empfinde lebhaft das Bedürfnis, etwas für diese Sache zu tun.

PAGE 63

Der Zionismus stellt wirklich ein neues jüdisches Ideal dar, das dem jüdischen Volk wieder Freude an seiner Existenz geben kann. . . . Ich bin sehr froh, Weizmanns Einladung Folge geleistet zu haben.

PAGE 63

Ich glaube, dass diese Sache nach und nach prächtig werden wird; mein jüdisches Heiligenherz frohlockt.

PAGE 63

Es ist nach meiner Ansicht ungerecht, die zionist. Bewegung als "nationalistisch" zu verurteilen. Theodor Herzl kam dadurch zu seiner Mission, dass er—vorher rein kosmopolitisch eingestellt—während des Dreyfus-Prozesses in Paris plötzlich mit aller Deutlichkeit fühlte, wie prekär die Situation der Juden in den westlichen Ländern war. Und er zog mutig die Konsequenzen. Wir werden zurückgesetzt oder erschlagen, nicht weil wir Deutsche, Franzosen, Amerikaner etc. "jüdischen Glaubens" sind, sondern einfach weil wir Juden sind. So ist es schon durch unsere äussere prekäre Lage bedingt, dass wir unabhängig von unserer Staatsangehörigkeit zusammenhalten müssen.

Der Zionismus hat den deutschen Juden keinen erheblichen Schutz gegen Vernichtung gegeben, wohl aber den Überlebenden die innere Kraft, den Zusammenbruch mit Würde zu überstehen, ohne das gesunde Selbstgefühl zu verlieren. Denken Sie daran, dass vielleicht Ihren Kindern ein ähnliches Schicksal beschieden sein mag!

PAGE 64

Ich danke Ihnen noch nachträglich dass Sie mir geholfen haben, mir die jüdische Seele zu Bewusstsein zu bringen.

PAGE 64

Auch wer der Relativität
Lehre doch nicht ganz versteht
Und wem die Koordinaten
Niemals was zuleide taten,
Auch für diesen doch ergibt sich,
Wenn man jung noch ist mit siebzig,
Dann beweist man damit faktisch
Jene Lehre doch auch praktisch.

Und wenn heute wohl in Mengen
Gratulanten Sie bedrängen,
Will auch ich mich nicht genieren
Uns und Ihnen gratulieren.
Als Unsrer gilt er ohne Fehl
Dem Volk vom Lande Israel.

PAGE 65

Worum andre sich oft grämen
Weisst du mit Humor zu nehmen
Denn du hast dir halt gedacht
Dass uns Gott so hat gemacht.

Der mit Unrecht sich tat rächen
Er, der selber schuf die Schwächen
Denen wehrlos wir erliegen
Ob in Elend, ob in Siegen.

Statt mit sturem Ernst zu richten
Bringst Erlösung durch dein Dichten
Dass die Frevler und die Frommen
All auf ihre Rechnung kommen.

PAGE 66

Ich kann mir keinen persönlichen Gott denken, der die Handlungen der einzelnen Geschöpfe direkt beeinflusste oder über seine Kreaturen direkt zu Gericht sässe. Ich kann es nicht, trotzdem die mechanistische Kausalität von der modernen Wissenschaft bis zu einem gewissen Grade in Zweifel gestellt wird. Meine Religiosität besteht in einer demütigen Bewunderung des unendlich überlegenen Geistes, der sich in dem Wenigen offenbart, was wir mit unserer schwachen und hinfälligen Vernunft von der Wirklichkeit zu erkennen vermögen. Moral ist eine höchst wichtige Sache, aber für uns, nicht für Gott.

PAGE 67

Vom skeptischen Empirismus etwa Mach'scher Art herkommend hat das Gravitationsproblem mich zu einem gläubigen Rationalisten gemacht, d.h. zu einem, der die einzige zuverlässige Quelle der Wahrheit in der mathematischen Einfachheit sucht.

PAGE 68

Ich habe mich mit diesem Grundproblem der Elektrizität nun mehr als 20 Jahre geplagt und bin recht mutlos geworden, ohne davon loskommen zu können. Ich bin überzeugt, dass eine ganz neue Erleuchtung kommen muss und glaube andererseits, dass die Flucht in die Statistik nur als vorübergehender Ausweg anzusehen ist, der an dem Wesentlichen vorbei geht.

PAGE 68

Sie sind der einzige mir bekannte Mensch, der dieselbe Einstellung zur Physik hat wie ich: Glaube an Erfassbarkeit der Realität durch etwas logisch Einfaches und Einheitliches. . . . Es scheint hart, dem Herrgott in seine Karten zu gucken. Aber dass er würfelt und sich "telepathischer" Mittel bedient (wie es ihm von der gegenwärtigen Quantentheorie zugemutet wird) kann ich keinen Augenblick glauben.

PAGE 69

Je mehr man den Quanten nachjagt, desto besser verbergen sie sich.

PAGE 69

Ich glaube nicht, dass die Grundgedanken der Relativitätstheorie in anderer Weise Beziehungen zur religiösen Sphäre beanspruchen können als die wissenschaftliche Erkenntnis überhaupt. Diese Beziehung sehe ich darin, dass tiefe Zusammenhänge in der objektiven Welt durch logisch einfache Gedanken erfasst werden können. Dies ist allerdings in der Relativitätstheorie in besonders vollkommenem Masse der Fall.

Das religiöse Gefühl, welches durch das Erlebnis der

logischen Fassbarkeit tiefliegender Zusammenhänge ausgelöst wird, ist von etwas anderer Art als dasjenige Gefühl, welches man gewöhnlich als religiös bezeichnet. Es ist mehr ein Gefühl der Ehrfurcht für die in den Dingen sich manifestierende Vernunft als solcher, welches nicht zu dem Schritte führt, eine göttliche Person nach unserem Ebenbilde zu formen—eine Person, die an uns Forderungen stellt und an unserem individuellen Sein Interesse nimmt. Es gibt darin weder einen Willen noch ein Ziel noch ein Soll, sondern nur ein Sein.

Deshalb sieht unsereiner in dem Moralischen eine rein menschliche Angelegenheit—aber allerdings die wichtigste in der menschlichen Sphäre.

PAGE 70

Unsere Zeit ist ausgezeichnet durch wunderbare Leistungen auf den Gebieten des wissenschaftlichen Verstehens und der technischen Anwendung der gewonnenen Einsichten; wer würde sich dessen nicht freuen? Indessen dürfen wir nicht vergessen, dass Wissen und Können allein die Menschen nicht zu einem würdigen und glücklichen Leben zu führen vermag. Die Menschheit hat allen Grund dazu, die Verkünder hoher moralischer Normen und Werte höher zu stellen als die Entdecker objektiver Wahrheit. Was die Menschheit Persönlichkeiten wie Buddha, Moses und Jesus verdankt, steht mir höher als alle Leistungen des forschenden und konstruktiven Geistes.

Die Gaben dieser Begnadeten müssen wir hüten und mit all unseren Kräften lebendig zu erhalten suchen, wenn das Menschengeschlecht nicht seine Würde, die Sicherheit seiner Existenz und die Freude am Dasein verlieren soll.

PAGE 71

Brüstet Euch nicht mit den paar bedeutenden Männern, die im Laufe der Jahrhunderte ohne Euer Verdienst auf Eurer Erde geboren wurden. Denket lieber darüber nach, wie Ihr sie jeweilen behandelt und wie Ihr ihre Lehren befolgt habt.

PAGE 71

Ich bin nicht Ihrer Meinung. Ich habe mein eigenes Leben stets für interessant und lebenswert gehalten und glaube fest daran, dass Möglichkeit und Aussicht dafür besteht, das Leben der Menschen im allgemeinen lebenswert zu gestalten. Die objektiven und psychologischen Möglichkeiten dafür scheinen mir durchaus gegeben zu sein.

PAGE 72

Ich bin tief erschüttert von der Nachricht über den furchtbaren Schlag, der über Sie beide so plötzlich und unerwartet hereingebrochen ist. Es ist das Schwerste, was älteren Menschen begegnen kann, und es ist kein Trost, dass so ungezählte Tausende von solchem Schicksal heimgesucht werden. Ich wage keinen Versuch Sie zu trösten, aber es drängt mich, Ihnen zu sagen, wie tief und schmerzlich ich mit Ihnen fühle und mit mir alle, die Ihre Herzsgüte kennengelernt haben.

Wir Menschen leben gewöhnlich in der Illusion einer Sicherheit und eines Zuhause-Seins in einer vertraut anmutenden physischen und menschlichen Umgebung. Wenn aber der Gang des Alltäglichen, Erwarteten unterbrochen wird, bemerken wir, dass wir sind wie Schiffbrüchige, die im offenen Meer auf einer elenden Planke balanzieren und vergessen haben, woher sie kommen, und nicht wissen, wohin sie treiben. Wenn man sich aber einmal in diese Erkenntnis wirklich hineingefunden hat, lebt sichs leichter, und es gibt keine einzige Enttäuschung mehr.

Hoffend, dass sich die Planken, auf denen wir schwimmen, bald wieder begegnen, grüsst Sie herzlich. . . .

PAGE 73

Man ist in eine Büffelherde geboren und muss froh sein, wenn man nicht vorzeitig zertrampelt wird.

PAGE 73

Sehr geehrter Herr:

Sie haben mir durch Ihre zarte Aufmerksamkeit eine grosse Freude bereitet. Die Benennung ist insofern treffend, als man nicht nur das Pflänzchen sondern auch mich auf dem ätherischen Gipfel nicht in Ruhe gelassen hat.

In dankbarer Anerkennung Ihrer sympathischen Geste bin ich. . . .

PAGE 73

Wo ich geh und wo ich steh
Stets ein Bild von mir ich seh,
Auf dem Schreibtisch, an der Wand
Um den Hals am schwarzen Band.

Männlein, Weiblein wundersam
Holen sich ein Autogramm,
Jeder muss ein Kritzel haben
Von dem hochgelehrten Knaben.

Manchmal frag in all dem Glück
Ich im lichten Augenblick:
Bist verrückt du etwa selber
Oder sind die andern Kälber?

PAGE 75

Was ich zu Bach's Lebenswerk zu sagen habe: Hören, spielen, lieben, verehren und—das Maul halten.

PAGE 75

Zu Schubert habe ich nur zu bemerken: Musizieren, Lieben—und Maulhalten!

PAGE 76

(1) Bach, Mozart und einige alte Italiener und Engländer sind meine Lieblinge in der Musik. Beethoven erheblich weniger, wohl aber Schubert.

(2) Ob mir Bach oder Mozart mehr bedeutet kann ich unmöglich sagen. Ich suche in der Musik keine Logik, sondern bin überhaupt ganz unbewust, kenne keine Theorien. Nie gefällt mir ein Werk, dessen innere Einheit ich nicht gefühlsmässig erfassen kann. (Architektur)

(3) Händel empfinde ich immer gut, ja vollkommen, aber von einer gewissen Flachheit. Beethoven ist mir zu dramatisch und zu persönlich.

(4) Schubert ist mir einer der liebsten wegen seines ungeheuer vollkommenen Gefühlsausdruckes und gewaltigen melodiösen Erfindungskraft. Bei den grösseren Werken stört mich aber ein gewisser Mangel an Architektonik.

(5) Schumann ist für mich in den kleinen Dingen reizvoll durch Originalität und Gefühlsreichtum, aber der Mangel an formaler Grösse lässt mich nicht zur vollkommenen Freude kommen. Bei Mendelssohn empfinde ich bedeutende formale Begabung, aber ein undefinierbarer Mangel an Tiefe, der oft bis zur Banalität geht.

(6) Von Brahms finde ich ein paar Lieder und Kammermusik-Werke wirklich bedeutend, auch im Aufbau. Die meisten Werke aber haben für mich keine innere Überzeugungskraft. Ich begreife nicht, dass es notwendig war, sie zu schreiben.

(7) Wagners Erfindung bewundere ich, empfinde aber den Mangel an architektonischer Struktur als Decadenz. Ausserdem empfinde ich die musikalische Persönlichkeit als unbeschreiblich widerwärtig, sodass ich ihn meist nur mit Widerwillen anhören kann.

(8) Strauss empfinde ich als begabt, aber ohne innere Wahrhaftigkeit und nur auf äussere Wirkung bedacht. Ich kann nicht sagen, dass mir die moderne Musik über-

haupt gestohlen werden kann. Debussy empfinde ich als feinfarbig, aber strukturarm. Grosse Begeisterung kann ich für so etwas nicht aufbringen. . . .

PAGE 78

Ein wirklicher Meister kann nur einer sein, der sich einer Sache mit ganzer Kraft und ganzer Seele hingibt. Deshalb verlangt Meisterschaft einen ganzen Mann. Dies zeigt Toscanini in jeder seiner Lebensäusserungen.

PAGE 78

Die Musik *wirkt* nicht auf die Forschungsarbeit, sondern beide werden aus derselben Sehnsuchtsquelle gespeist und ergänzen sich bezüglich der durch sie gewährten Auslösung.

PAGE 79

Immer noch kämpfe ich mit denselben Problemen wie vor 10 Jahren. Kleines gelingt, aber das eigentliche Ziel bleibt mir unerreichbar, wenn es auch manchmal in greifbare Nähe gerückt scheint. Es ist hart aber doch beglückend, hart, weil das Ziel zu gross ist für meine Kräfte, aber beglückend, weil es immunisiert gegen die Zwischenfälle des persönlichen Daseins.

In die Menschenwelt hier lebe ich mich nicht mehr hinein, dazu war ich schon zu alt, als ich herkam und—es war eigentlich in Berlin und früher in der Schweiz auch nicht anders. Zum Einspänner ist man schon geboren. Sie verstehen das, weil Sie auch so einer sind.

PAGE 80

Ich preise mich glücklich, Euch nach so vielen Jahren hier begrüssen zu dürfen. Ich hatte mir Schweigen auferlegt, weil jede Zeile von mir nach Barbarien den Adressaten gefährdet. Ihr lieber Schopenhauer hat einmal gesagt, dass die Menschen in ihrem Jammer es nicht zur Tragödie bringen, sondern dass sie dazu verdammt sind, in der Tragikomödie stecken zu bleiben. Wie wahr das ist, und wie oft habe ich diesen Eindruck erlebt. Gestern vergöttert, heute gehasst und angespuckt, morgen vergessen und übermorgen zum Heiligen avanciert. Nur der Humor rettet über alles hinweg, den wollen wir festhalten, solange der Schnaufer dauert.

PAGE 80

Ich war sehr gerührt über Ihre lieben Zeilen und gratuliere herzlich nachträglich. Ich weiss, dass ich so viel Rühmens nicht im Entferntesten verdiene, freue mich aber über die herzliche Gesinnung, die aus Ihren Worten leuchtet.

Ich glaube, wir dürfen nun doch hoffen, zu erleben, dass von dem unaussprechlichen Unrecht etwas gesühnt wird. Aber all der Jammer, all die Verzweiflung, all die sinnlosen Vernichtungen von Menschenleben können nicht wieder gut gemacht werden. Wir dürfen aber doch hoffen, dass es auch den stumpfsten Geschöpfen eingehämmert werden wird, dass Lüge und Vergewaltigung nicht für die Dauer triumphieren können.

An Ihnen sieht man, was für einen unerschütterlichen Halt das Streben nach Wahrheit verleihen kann. Auch ich verdanke dieser Einstellung die einzige wahre Befriedigung. Man fühlt es, dass man in der zeitlosen Gemeinschaft der Menschen dieser Art eine Art Zuflucht hat, die keine Verzweiflung und kein Gefühl hoffnungsloser Vereinsamung aufkommen lässt.

PAGE 81

"Die Schlussworte des schönen Gebetes 'Der Herr hat gegeben, der Herr hat genommen, der Name des Ewigen sei gepriesen!'—bedeuten die Fülle des Lebens, das immer gibt und wieder nimmt, um abermals zu geben."

PAGE 81

Sie nehmen eine entschiedene Stellung ein bezüglich Hitlers Verantwortlichkeit. Ich habe eigentlich nie an diese feineren Unterscheidungen geglaubt, die durch Juristen den Medizinern aufgezwungen werden. Es gibt ja objektiv keine Willensfreiheit. Ich denke, dass wir uns gegen Menschen schützen müssen, die für andere eine Gefahr sind, ganz unabhängig davon, wie ihre Handlungen determiniert sein mögen. Was braucht es da das Kriterium der Verantwortlichkeit? Ich glaube, dass der erschreckende Verfall im ethischen Verhalten der Menschen in erster Linie mit der Mechanisierung und Entpersönlichung unseres Lebens zu tun hat—ein verhängnisvolles Nebenprodukt der Entwicklung des wissenschaftlich-technischen Geistes. Nostra culpa! Ich sehe nicht den Weg, um diesem verhängnisvollen Mangel beizukommen. Der Mensch erkaltet schneller als der Planet, auf dem er sitzt.

PAGE 82

Ich erfahre von verschiedenen Freunden, dass Sie (kaum möcht' ichs für möglich halten!) in diesen Tagen Ihren 80. Geburtstag feiern. Solche Menschen wie wir beide sterben zwar wie alle, aber sie werden nicht alt, solange sie leben. Ich meine damit, sie stehen immer noch neugierig wie Kinder vor dem grossen Rätsel, in das wir mitten hineingesetzt sind. Dies gibt eine Distanz gegenüber allem Unbefriedigenden in der menschlichen Sphäre—und das ist nicht wenig. Wenn ich morgens den Ekel über das bekomme, was einem die N.Y. Times auftischt, dann denke ich immer, dass es noch besser ist als die Hitlerei, der man mit knapper Mühe den Garaus gemacht hat.

PAGE 83

Das Studium und allgemein das Streben nach Wahrheit und Schönheit ist ein Gebiet, auf dem wir das ganze Leben lang Kinder bleiben dürfen.

Adrianna Enriques zum Andenken an die Bekanntschaft von Oktober 1921.

PAGE 83

Ich habe mich . . . zu der Überzeugung durchgearbeitet, dass die Abschaffung der Todesstrafe wünschenswert ist. Begründung:
1) Irreparabilität im Falle eines Justizirrtums,
2) Nachteiliger moralischer Einfluss der Hinrichtungsprozedur auf diejenigen, welche mit der Execution direkt oder indirekt zu tun haben.

PAGE 84

Was ich über Krieg und Todesstrafe denke? Das letztere ist einfacher. Ich bin überhaupt nicht für die Strafe, sondern nur für Massregeln im Dienste der Gesellschaft und deren Schutz. Im Prinzip wäre ich nicht dagegen, in diesem Sinne wertlose oder gar schädliche Individuen zu töten; ich bin nur deshalb dagegen, weil ich den Menschen, d.h. den Gerichten misstraue. Ich schätze nämlich am Leben mehr die Qualität als die Quantität, so wie sich in der Natur die Gesetzmässigkeit als höhere Realität gegenüber dem Einzelding darstellt.

PAGE 84

Was Ihre Frau gesagt hat, finde ich gar nicht schlecht. Es ist wahr, dass ein Mensch, der ein gewisses Renomee in der Öffentlichkeit hat, in geringerem Masse gefährdet ist als ein weniger bekannter Mensch. Wie aber sollte ein Mensch, den die Leute kennen, von seinem Namen einen besseren Gebrauch machen als dadurch, dass er von Zeit zu Zeit öffentlich ausspricht, was er für nötig hält? Der Vergleich mit Sokrates passt insofern nicht, als für letzteren Athen die Welt bedeutete, während ich mich nie mit einem besonderen Lande identifizierte, am wenigsten mit dem politischen Deutschland, mit dem mich eigentlich nichts verband als meine Stellung an der preussischen Akademie und die Sprache, die ich als Kind lernte.

Wenn ich nun auch überzeugter Demokrat bin, so weiss ich doch, dass die Menschenwelt stagnieren würde, wenn nicht eine Minorität wohlmeinender und aufrechter Menschen für ihre Überzeugung Opfer bringen würden. Gegenwärtig ist dies besonders nötig. Ich brauche das wohl nicht besonders zu begründen.

PAGE 85

Der wirkliche menschliche Fortschritt gründet sich weniger auf Erfindergehirne als auf das Gewissen solcher Männer wie Brandeis.

PAGE 86

In tiefer Verehrung und Sympathie drücke ich Ihnen die Hand bei Gelegenheit Ihres 80. Geburtstages. Ich kenne keinen Zweiten der so tiefe Geistesgaben mit so völliger Selbst-Entäusserung verbände, im stillen Dienste an der Gesamtheit den ganzen Sinn seines Daseins fände. Wir danken Ihnen alle nicht nur für das, was Sie geleistet und gewirkt haben, sondern wir fühlen uns auch beglückt dadurch, dass es einen solchen Mann überhaupt gibt in unserer an wirklichen Persönlichkeiten so armen Zeit.

Mit ehrerbietigen Grüssen. . . .

PAGE 87

Beim Lesen des White'schen Aufsatzes erlebt man es als überzeugende Wahrheit: Zu wahrer menschlicher Grösse gibt es nur einen Weg–den durch die Schule des Leidens. Wenn das Leiden aus der seelischen Blindheit und Stumpfheit einer traditionsgebundenen Gesellschaft entspringt, pflegt es die Schwachen zum blinden Hass zu degradieren, die Starken aber zu erheben zu einer moralischen Überlegenheit und magnanimity, die sonst dem Menschen kaum zugänglich ist.

Ich denke, dass jeder empfängliche Leser so wie ich selber, Walter White's Aufsatz mit einem Gefühl wahrer Dankbarkeit aus der Hand legen wird. Er hat uns an seinem schmerzvollen Wege zu menschlicher Grösse teilnehmen lassen durch eine schlichte biographische Erzählung, deren Überzeugungskraft unwiderstehlich ist.

PAGE 87

Ihr Vorschlag erscheint an sich vernünftig: Organisation der Wirtschaft durch eine kleine Zahl Menschen, die sich als productiv und als stark und selbstlos interessiert an einer Besserung der Verhältnisse erwiesen haben. Dagegen halte ich nichts von Ihrer Auswahl-Methode durch "tests". Es ist eine Ingenieur-Idee typischer Art, die Ihrem eigenen Ausspruch "Der Mensch ist keine Maschine" durchaus nicht gemäss ist.

Ferner bitte ich Sie, eines zu bedenken: Es genügt nicht, die zehn geeignetsten Personen ausfindig zu machen. Es muss auch bewirkt werden, dass die Völker sich ihren Anordnungen fügen. Wie dies zuweg gebracht werden soll, davon kann ich mir kein Bild machen. Diese Frage ist ungleich schwieriger als die der Auswahl geeigneter Persönlichkeiten. Denn schon ziemlich mittelmässige Leute könnten das Werk in passabler Weise vollbringen, verglichen mit den heute und überhaupt bisher bestehenden Verhältnissen. Bisher verdankten die Führenden ihre Macht in der Hauptsache nicht etwa der Fähigkeit selber zu denken und Entschlüsse zu fassen sondern andere zu überreden, auf sie Eindruck zu machen und ihre Schwächen auszunutzen.

Das alte Problem: Was muss man tun, um Fähigen und Wohlwollenden die Macht über die Menschen zu geben? hat bis jetzt allen Anstrengungen gespottet. Leider scheint es mir, dass auch Sie keinen Weg gefunden haben, es zu lösen.

PAGE 88

Wie ist es nur möglich, dass diese kulturliebende Zeit so grässlich amoralisch ist? Ich komme immer mehr dazu, alles andere gegen die Nächstenliebe und Menschenfreundlichkeit gering einzuschätzen. . . . Unser ganzer gepriesener Fortschritt der Technik, überhaupt die Civilisation, ist der Axt in der Hand des pathologischen Verbrechers vergleichbar.

PAGE 89

Wenn ich nun darüber nachdenke, was eigentlich Toleranz sei, fällt mir die drollige Definition ein, die der humorvolle Wilhelm Busch von der Enthaltsamkeit gegeben hat:

Enthaltsamkeit ist das Vergnügen
An Dingen, welche wir nicht kriegen.

So möchte ich sagen: Toleranz ist das menschenfreundliche Verständnis für Eigenschaften, Auffassungen und Handlungen anderer Individuen, die der eigenen Gewohnheit, der eigenen Überzeugung und dem eigenen Geschmack fremd sind. Tolerant heisst also nicht Gleichgültigkeit gegen das Handeln und Fühlen des oder der andern; es muss auch Verständnis und Einfühlung dabei sein. . . .

Das Grosse und Edle kommt von der einsamen Persönlichkeit, sei es ein Kunstwerk oder eine bedeutende schöpferische wissenschaftliche Leistung. Die europäische Kultur erlebte ihren wichtigsten Aufschwung aus dumpfem Verharren heraus, als die Renaissance dem Individuum Möglichkeiten zur freien Entfaltung bot.

Die wichtigste Art der Toleranz ist deshalb die der Gesellschaft und des Staates gegen das Individuum. Der Staat ist gewiss nötig, um dem Individuum die Sicherheit für seine Entwicklung zu geben, aber wenn er zur Hauptsache wird und der einzelne Mensch zu seinem willenlosen Werkzeug, dann gehen alle feineren Werte verloren. Wie der Fels erst verwittern muss, damit Bäume auf ihm wachsen können, und der Ackerboden erst aufgelockert werden muss, damit er seine Fruchtbarkeit entfalten kann, so spriessen auch aus der menschlichen Gesellschaft nur dann wertvolle Leistungen hervor, wenn sie genügend gelockert ist, um dem einzelnen freie Entfaltung seiner Fähigkeiten zu ermöglichen.

PAGE 91

Als Gott der Allmächtige seine ewigen Naturgesetze aufstellte, da plagte Ihn ein Bedenken, das Er auch in der Folgezeit nicht zu überwinden vermochte: Was für eine peinliche Situation würde entstehen, wenn die hohen Autoritäten des dialektischen Materialismus später einige dieser Gesetze oder gar all miteinander als ungesetzlich erklärten?

Als Er dann später dazu überging, die Propheten und Weisen des dialektischen Materialismus zu erschaffen, da schlich ein einigermassen ähnliches Bedenken in Seine Seele. Aber bald beruhigte Er sich, indem Er glaubte, darauf vertrauen zu dürfen, dass diese Propheten und Weisen niemals zu dem Schluss kommen würden, dass die Lehren des dialektischen Materialismus der Vernunft und Wahrheit widersprächen.

PAGE 92

Weder auf meinem Sterbebette noch vorher werde ich mir eine solche Frage vorlegen. Die Natur ist kein Ingenieur oder Unternehmer, und ich bin selber ein Stück Natur.

PAGE 95

Das Streben nach moralischem Handeln ist das wichtigste Streben der Menschen. Sein inneres Gleichgewicht, ja, seine Existenz hängen davon ab. Moralisches Handeln allein kann dem Leben Schönheit und Würde verleihen.

Dies den Jungen lebendig zu machen und zu voller Klarheit zu bringen ist wohl die Hauptaufgabe der Erziehung. Die Gestaltung des moralischen Ideals sollte nicht an einen Mythos gebunden und mit einer Autorität verknüpft werden, dass nicht durch Zweifel an dem Mythos oder der Berechtigung der Autorität das Fundament des richtigen Urteilens und Handelns gefährdet werde.

PAGE 96

Wenn die Bekenner der gegenwärtigen Religionen sich ernstlich bemühen würden, im Geiste der Begründer dieser Religionen zu denken, zu urteilen und zu handeln, dann würde keine auf den Glauben gegründete Feindschaft zwischen den Bekennern verschiedener Religionen existieren. Noch mehr, sogar die Gegensätze im Glauben würden sich als unwesentlich herausstellen.

PAGE 96

Die feierliche Versammlung des heutigen Tages hat eine tiefe Bedeutung. Wenige Jahre nur trennen uns von dem furchtbarsten Massenverbrechen, das die moderne Geschichte aufzuweisen hat, ein Verbrechen, nicht begangen von einem fanatischen Haufen, sondern in kalter Berechnung. Das Schicksal der überlebenden Opfer der deutschen Verfolgung legt Zeugnis davon ab, wie schwach das moralische Gewissen der Menschheit geworden ist.

Die heutige Versammlung legt Zeugnis davon ab, dass die besseren Menschen nicht Willens sind, das Furchtbare schweigend hinzunehmen. Diese Versammlung ist beseelt von dem Willen, dem menschlichen Individuum seine Würde und natürlichen Rechte zu sichern. Sie steht ein für die Erkenntnis, dass ein erträgliches Dasein für die Menschen—ja eben das nackte Dasein—an das Festhalten an den ewigen moralischen Forderungen gebunden ist.

Für diese Haltung spreche ich ihr als Mensch und als Jude Anerkennung und Dank aus.

PAGE 99

Liebe Gertrud,

Vor mir liegt das niedliche Lineal, das Sie mir gesandt haben. Bisher wars der Intuition überlassen, ob etwas bei mir gerade oder krumm, parallel oder schief herauskam. Ich sehe aber, dass Sie es lieber vermeiden wollen, in Gottes Hand zu sein, wenn man es vermeiden kann (so deute ich das Lineal.).

PAGE 99

Sie haben mir mit dem Faraday-Büchlein eine grosse Freude gemacht. Dieser Mann liebte die rätselhafte Natur wie ein Liebhaber die ferne Geliebte. Es gab noch nicht das öde Spezialistentum, das mit Hornbrille und Dünkel die Poesie zerstört.

PAGE 100

Unbehaglich macht mich stets das Wörtchen "wir"
Denn man ist nicht eins mit einem andern Tier.
Hinter allem Einverständnis steckt
Stets ein Abgrund, der noch zugedeckt.

PAGE 101

Glauben Sie wirklich, dass Kaiser Karl so begeistert gewesen wäre, wenn Tizian von ihm eine Ansichtspostkarte hergestellt hätte, die jeder Piefke für zehn Pfennige geliefert bekommt? Ich glaube, er hätte zwar Tizian nicht minder freudig den Pinsel aufgehoben, hätte ihn aber doch gebeten, ihn mit solcher Publizität zu verschonen, wenigstens bei Lebzeiten.

Seien Sie also nicht böse, wenn auch ich in dieser Weise empfinde. Ich muss übrigens in wenigen Tagen nach Kalifornien verreisen und habe alle Hände voll zu tun.

P.S. Heiliger St. Florian, verschon' mein Haus, zünd andere an!

PAGE 101

Alter und Krankheit machen es mir unmöglich, mich bei solchen Gelegenheiten zu beteiligen, und ich muss auch gestehen, dass diese göttliche Fügung für mich etwas Befreiendes hat. Denn alles was irgenwie mit Personenkultus zu tun hat, ist mir immer peinlich gewesen.

PAGE 102

Es ist eigentlich rätselhaft, was einen antreibt, die Arbeit so verteufelt ernst zu nehmen. Für wen? Für sich? –Man geht doch bald. Für die Mitwelt? Für die Nachwelt? *Nein*, es bleibt rätselhaft.

PAGE 103

Jeder zeiget sich mir heute
Von der allerbesten Seite
Und von fern und nah die Lieben
Haben rührend mir geschrieben
Und mit allem mich beschenkt
Was sich so ein Schlemmer denkt-
Was für den bejahrten Mann
Noch in Frage kommen kann.
Alles naht mit süssen Tönen
Um den Tag mir zu verschönen.
Selbst die Schnorrer ohne Zahl
Widmen mir ihr Madrigal.

Drum gehoben fühl' ich mich
Wie der stolze Adlerich.
Nun der Tag sich naht dem End'
Mach ich Euch mein Kompliment,
Alles habt Ihr gut gemacht
Und die liebe Sonne lacht.

A. Einstein
peccavit 14.III. 29.

PAGE 104

". . . In einigen hundert Jahren wird der gemeine Mann unsere Zeit als die Periode des Weltkrieges kennen, aber der Gebildete wird das erste Viertel des Jahrhunderts mit Ihrem Namen verbinden, so wie sich heute beim Ausgang des 17. Jahrhunderts die einen an die Kriege Ludwig des Vierzehnten und die andern an Isaac Newton erinnern. . . ."

PAGE 104

". . . Nun, jetzt bist Du ja geborgen, jenem Hass unerreichbar; Du bist, wie ich Dich kenne, auch innerlich mit ihm fertig geworden und stehst *über* Deinem Schicksal. Dein Werk aber ist und bleibt erst recht unerreichbar aller Leidenschaft, und es dauert, solange es eine Kulturmenschheit auf Erden gibt. . . ."

PAGE 105

Liebe Nachwelt!

Wenn Ihr nicht gerechter, friedlicher und überhaupt vernünftiger sein werdet, als wir sind, bezw. gewesen sind, so soll euch der Teufel holen.

Diesen frommen Wunsch mit aller Hochachtung geäussert habend bin ich euer (ehemaliger)

gez. Albert Einstein

PAGE 106

Die Philosophie gleicht einer Mutter, die alle übrigen Wissenschaften geboren und ausgestattet hat. Man darf sie in ihrer Nacktheit und Armut daher nicht geringschätzen, sondern muss hoffen, dass etwas von ihrem Don-Quichote-Ideal auch in ihren Kindern lebendig bleibe, damit sie nicht in Banausentum verkommen.

PAGE 110

Ein bei aller Tragik groteskes Schauspiel ist es, dass sich im Herzen Europas abspielt, Deutschland zu ewiger Schande und auch recht unrühmlich für die Gemeinschaft der Nationen, welche sich selber als "Kulturmenschheit" bezeichnet.

Das deutsche Volk ist durch Jahrhunderte hindurch von einer sich ewig erneuernden Schar von Schulmeistern und Unteroffizieren sowohl zu emsiger Arbeit und mancherlei Wissen als auch zu sklavischer Unterwürfigkeit und zu militärischem Drill und Grausamkeit erzogen worden. Die republikanisch-demokratische Verfassung der Nachkriegszeit passte zu diesem Volke etwa wie der Anzug des korpulenten Onkels für den kleinen Hans. Dazu kamen dann die Inflation und die bösen Krisenjahre, sodass jeder Mensch in Furcht und Spannung lebte.

Da erschien in Hitler einer von den Armen im Geiste, unbrauchbar für jegliche Arbeit, erfüllt von Neid und Erbitterung gegen alle, die von Natur und Schicksal mehr begünstigt erschienen als er. Er war der Sohn einer Kleinbürger-Familie mit gerade genug Klassendünkel, um auch die um einen vernünftigen Ausgleich der Lebensbedingungen kämpfenden Arbeiter zu hassen. Am meisten aber hasste er Geist und Bildung, die ihm am unerbittlichsten versagten Gaben. In seiner verzweifelten Lage fand er heraus, dass seine verworrenen, von Hass getragenen Reden stürmischen Anklang fanden bei den ach so Zahlreichen, deren Lage und Gesinnung der seinigen ähnlich waren. Solche verzweifelten Existenzen las er auf der Strasse und im Wirtshause auf, scharte sie um sich. So wurde er zum Politiker.

Was ihn aber zum "Führer" prädestinierte, war sein bitterer Hass gegen das Ausland und gegen eine wehrlose Minorität, die deutschen Juden. Diese hasste er besonders als die Vertreter einer ihm unheimlichen Geistigkeit, die er nicht ganz mit Unrecht als undeutsch empfand.

Unablässige Hassreden gegen diese beiden Arten von "Feinden" eroberte die Massen, denen er glänzenden Sieg und ein goldenes Zeitalter versprach. Auch nützte er weidlich den gewohnheitsmässigen Drang der seit Jahrhunderten militärisch gedrillten Massen zum Marschieren, blinden Gehorsam, Befehlen und zur Menschenschinderei für seine Zwecke aus. So wurde er zum "Führer".

Geld floss ihm reichlich zu, nicht zuletzt von der besitzenden Klasse, die in ihm ein Werkzeug sah zur Verhinderung der in der Zeit der Republik in gewissem Umfange begonnenen sozialen und wirtschaftlichen Befreiung des Volkes. Dem Volk aber schmeichelte er durch jene romantischen Phrasen der Vaterländerei, an die es von der Vorkriegszeit her gewöhnt war, sowie durch jenen Schwindel von der Überlegenheit einer von den Antisemiten zu ihren besonderen Zwecken erfundenen "arischen" beziehungsweise "nordischen" Rasse. Die Verworrenheit seines Geistes macht es mir unmöglich zu beurteilen, bis zu welchem Grade er selbst an den Unsinn glaubte, den er unablässig predigte. Die aber, welche sich um ihn scharten und sich durch die Woge nach oben treiben liessen, waren meist verworfene Zyniker, die sich der Verlogenheit ihrer Mittel voll bewusst waren.

PAGE 112

Was dieser Mann den in Deutschland gefangenen und dem sicheren Untergang entgegenstehenden Brüdern gewesen ist, kann der in dem Gefühl äusserer Sicherheit Dahinlebende nicht voll begreifen. Er empfand es als selbstverständliche Pflicht, in dem Lande ruchloser Verfolgung auszuharren, um seinen Brüdern bis zuletzt eine seelische Stütze zu sein. Keine Gefahr scheuend, unterhandelte er mit den Vertretern einer aus ruchlosen Mördern bestehenden Regierung und wahrte in jeder Situation seine und seines Volkes Würde.

PAGE 113

In dem Wunsche, Ihrem schönen Unternehmen zu dienen und doch unfähig, einen Beitrag auf dem Gebiete unseres verehrten und geliebten Freundes zu liefern, bin ich auf die bizarre Idee verfallen, etwas aus der eigenen Lebenserfahrung, was unseren Freund ein bischen ergötzen könnte, in Pillenform zusammen zu stellen, wobei allerdings nur die erste Pille direkte Beziehung zu ihm beanspruchen darf.

PAGE 113

Um ein tadelloses Mitglied einer Schafherde sein zu können, muss man vor allem ein Schaf sein.

PAGE 113

Heil dem Manne, der stets helfend durchs Leben ging, keine Furcht kannte und dem jede Aggressivität und jedes Ressentiment fremd war. Von solchem Holz sind die Idealgestalten geschnitzt, die der Menschheit Trost bieten in den Situationen selbstgeschaffenen Leidens.

PAGE 114

In Tagen, in denen ein sittlicher Zweifel nur ein Nein zu finden schien oder diese Frage nach dem Menschlichen selbst unterhalb des Zweifels blieb, habe ich an Sie denken dürfen, und eine Ruhe, eine Bejahung kam über mich. In so mancher Stunde haben Sie vor mir gestanden und zu mir gesprochen.

PAGE 114

Man hat das Gefühl, dass so ein Mensch für immer weiterlebt, wie ja ein Mensch wie Beethoven auch niemals sterben kann. Nur ist eben das Unersetzliche, dass gerade seine Lebendigkeit so stark zu seinem Wesen gehörte und man sich so schwer vorstellen kann, dass dieser so unsagbare bescheidene und schlichte Mensch nicht mehr unter uns weilen soll. Er war sich seines einzigartigen Schicksals bewusst und wusste um seine Grösse. Aber gerade weil diese Grösse so überragend war, hat sie ihn bescheiden und demütig gemacht, und zwar nicht aus Pose, sondern aus innerem Bedürfnis.

PAGE 115

Ich bin der Mann, dem Sie über die belgische Akademie geschrieben haben. . . . Lesen Sie keine Zeitung, suchen Sie ein paar Gesinnungsgenossen und lesen Sie die wunderbaren Schriftsteller früherer Zeiten, Kant, Goethe, Lessing und die Klassiker des Auslands und erfreuen Sie sich an der wundervollen Natur der Münchner Gegend. Denken Sie unablässig, dass Sie gewissermassen unter fremden Geschöpfen auf dem Mars leben und tilgen Sie jede tiefere Anteilnahme an dem Treiben dieser Geschöpfe. Schliessen Sie Freundschaft mit ein paar Tieren. Dann werden Sie wieder ein froher Mensch werden und nichts kann Sie anfechten. Bedenken Sie dass die feineren und edleren Menschen stets allein sind und sein müssen und dass sie dafür die Reinheit ihrer eigenen Atmosphäre geniessen dürfen.

Es grüsst Sie kameradschaftlich und drückt Ihnen bewegt die Hand Ihr.

E.

Einstein: a Brief Chronology

ALBERT EINSTEIN was born in Ulm, Germany, on 14 March 1879, his sister Maja being born in Munich two and a half years later. Given a magnetic compass at the age of five, he was overwhelmed by a feeling of awe and wonder that remained with him for the rest of his life and underlay his greatest scientific achievements. At age twelve he felt a similar wonder on first looking into a geometry textbook.

He hated the discipline and rote learning in the German schools and at age fifteen became a school dropout. In 1896 he entered the Polytechnic Institute in Zurich, Switzerland. On graduating in 1900, having antagonized his professors, he was unable to obtain an academic position.

In 1901 he became a Swiss citizen. In 1902, after many discouragements, he obtained a job in the Swiss patent office in Bern. He then married a former fellow student, Mileva Maric, by whom he had two sons; but the marriage ended in amicable divorce in 1919.

Meanwhile, at the patent office, in the fabulous year 1905, his genius burst into dazzling flower. The theory of relativity was only one of his several major contributions in that year.

Not till 1909 did he leave the patent office, but then advancement was rapid and by 1914 he was at the top of his profession as a salaried member of the Royal Prussian Academy of Sciences in Berlin.

As a Swiss citizen, he took no part in World War I, which broke out in August 1914. In 1915 he set forth his masterpiece, the general theory of relativity. In 1919 he married a widowed cousin, Elsa, who had two daughters by her first marriage. Later in 1919, with verification of a prediction of his theory, Einstein became world famous overnight. He was awarded the 1921 Nobel Prize in Physics.

The rest may be covered here in less detail since it hinges on one key year, 1933. In Germany Einstein's fame and brave outspokenness led to anti-Semitic attacks on both him and his theories. When the Nazis seized power early in 1933, he was in the United States. He never returned to Germany. Instead, he stayed a few months in Le Coq, in Belgium, spent a brief time in England, and in October of 1933 moved to the United States to join the faculty of the newly-founded Institute for Advanced Study in Princeton, New Jersey, where he remained for the rest of his life. He died on 18 April 1955.

Acknowledgments

WE are grateful to Herbert S. Bailey, Jr., Director of Princeton University Press, for the personal interest he has shown in this book from the time of its conception. His enthusiasm, encouragement, and advice lightened the way to its completion.

Our thanks go, also, to Sonja Bargmann, who subjected the translations to a searching scrutiny, not merely correcting errors but also making major improvements in the wording. We are grateful, too, to Gail Filion for her perceptive suggestions and editorial wisdom.

We thank: Mrs. Marianne Dreyfuss for permission to quote from a letter written by her grandfather Leo Baeck; L. F. Haber for permission to quote from a letter written by Fritz Haber; Elmar Lanczos for permission to quote from a letter written by Cornelius Lanczos; Theodor von Laue for permission to quote from a letter written by Max von Laue; J. Peemans, Chef-Adjoint du Cabinet du Roi, for permission to quote a poem written by the late Queen Elizabeth of Belgium; and, of course, Otto Nathan for permission to quote voluminously from writings of Einstein, and also for his lively interest in our book.

Library of Congress Cataloging in Publication Data

Main entry under title:

Albert Einstein, the human side.

English and German.
1. Einstein, Albert, 1879-1955. 2. Physicists–Biography 3. Physicists–Correspondence. I. Dukas, Helen. II. Hoffmann, Banesh, 1906-
QC16.E5A65 530'.092'4 [B] 78-70289
ISBN 0-691-08231-6
ISBN 0-691-02368-9 pbk.